Evapotranspiration - From Measurements to Agricultural and Environmental Applications

Evapotranspiration - From Measurements to Agricultural and Environmental Applications

Editor

Stefanos Palmer

scitus academics

Evapotranspiration - From Measurements to Agricultural and Environmental Applications
Edited by **Stefanos Palmer**

ISBN: 978-1-68117-241-5
Library of Congress Control Number: 2016934781

www.scitusacademics.com
Box No. 4766, 616 Corporate Way,
Suite 2, Valley Cottage,
NY 10989

Printed in United States of America on Acid Free Paper

Preface

Evapotranspiration (ET) is the sum of evaporation from the earth's surface and the ocean into the atmosphere. Evaporation is responsible for moving water from sources like soil, canopy and water into the air. Transpiration involves vaporizing liquid water in plant tissue and removing the atmospheric vapor. Crops primarily lose their water through stomata. These are small openings on the plant leaf through which pass water and gases. Water and some nutrients are removed and transported through the plant from the roots. Vaporization takes place in the leaf, in intercellular spaces, and the atmospheric exchange of vapors is controlled by the aperture of the stomach. Almost all water is lost by sweating and only a small fraction of water is used in the plant. Energy is needed to change the state of the water molecules from liquid to vapor. Direct solar radiation and ambient air temperature supply this energy to a lesser extent. The driving force for removing water vapor from the evaporating surface is the difference between water vapor pressure on the evaporating surface and environmental pressure. As evaporation progresses, the surrounding air gradually saturates and the process slows down and can be stopped when the wet air is not transferred to the atmosphere. The replacement of saturated air with drier air depends greatly on wind speed. In the evaluation of the evaporation process, solar radiation, air temperature, air humidity and wind speed are therefore considered climate parameters.

This book provides an overview of direct evapotranspiration measurement techniques with related applications for optimizing the use of water in agricultural practices and ecosystem research. Detailed measurements of the transpiration requirements of crops and different cultivars and the results of different irrigation schemes and techniques in semi-arid areas for open fields, greenhouse crops and potted crops are presented to optimize water use in agriculture. The book provides wide-ranging information for researchers and students who expect to address these issues.

Table of Contents

Chapter 1 Climate Change Effects on Evapotranspiration in Mexico 1

Chapter 2 Field Measurement of Cotton Seedling Evapotranspiration 17

Chapter 3 Recent Updates of the Calibration-Free Evapotranspiration Mapping (CREMAP) Method 41

Chapter 4 Estimation of Evapotranspiration and Crop Coefficients of Tendone Vineyards Using Multi-Sensor Remote Sensing Data in a Mediterranean Environment 61

Chapter 5 Influence of Potential Evapotranspiration on the Water Balance of Sugarcane Fields in Maui, Hawaii 89

Chapter 6 Evapotranspiration Characteristics of a Lowland Dry Evergreen Forest in Central Cambodia Examined Using a Multilayer Model 113

Chapter 7 Determination of Evapotranspiration and Water Use Efficiency in Crop Production 135

Chapter 8 Upflow Evapotranspiration System for the Treatment of On-Site Wastewater Effluent 153

Chapter 9 Monitoring of Evapotranspiration in a Semi-Arid Inland River Basin by Combining Microwave and Optical Remote Sensing Observations 181

Chapter 10 **Use of Evapotranspiration (ET) Landfill Covers to Reduce Methane Emissions from Municipal Solid Waste Landfills 221**

Chapter 11 **A Coupled Remote Sensing and Simplified Surface Energy Balance Approach to Estimate Actual Evapotranspiration from Irrigated Fields 237**

Index 263

Chapter 1

Climate Change Effects on Evapotranspiration in Mexico

Martín Mundo-Molina

Investigation Centre, Faculty of Engineer, Autonomous University of Chiapas, Tuxtla Gutiérrez, Mexico

ABSTRACT

In Mexico's case, the scarcity of water in the north of the country is worrying, even more from the agricultural point of view because according to the results of general circulation models, a major impact due to global warming is expected in this region, and it will have important repercussions on the rural sector in the north of Mexico. According to [1] the results of the sensitivity analysis made in this study indicate that the most vulnerable zone is the north of Mexico, wherein the increase of evapotraspiration (ET) is greater in comparison to the rest of the country; up to 8% annual average for a +3°C growth in mean annual temperature. Due to some limitations in this preliminary investigation (e.g., global temperature data was used without regionalizing it), it was decided to make more detailed studies to estimate the climate change effects on ET on a regional scale, using the downscaling method to adjust temperature data. In this study a new methodology to estimate the ET before climate change scenarios is introduced, which includes the selection of the Hargreaves-Samani method (HS), calibrated and compared against the Penman-Monteith ASCE method in various irrigation districts in the northern part of the country, obtaining ET estimations with a 93% precision. This procedure was applied to nine states in north Mexico: Baja California, Baja California Sur, Chihuahua, Sinaloa, Sonora, Tamaulipas, Nuevo León, Coahuila y Durango. The principal results are enunciated as follows: the ET variations between the contemporary scenario and the 2030 scenario are quite significant, according to the data of 160 meteorological stations; for temperature variations between 0.1°C to 0.45°C the corresponding ET fluctuation goes from 2% in the current scenario to 7% in the 2030 scenario. These obtained percentages are greater than the ones expected to happen for the precision of the method. It is important to note that a 7% rise of ET (related to a regional temperature increase of approximately one degree) would represent in practice having more millions of m3 of water in dams to satisfy the water demand of crops.

Keywords: Climatic Change, Reference Evapotranspiration, Irrigation Agriculture

1. INTRODUCTION

The observation networks of the World Meteorological Organization (WMO) have registered a raise in worldwide mean temperature from 3°C to 6°C during the last hundred years because of increment of greenhouse gases in the atmosphere during the last 2000 years.

According to stored data and physical evidence, the increase of greenhouse gases has originated Earth's over- heating. In 1998, the United Nations Environment Programme (UNEP) and WMO created the Intergovernmental Panel on Climate Change (IPCC). The objective of this group is to analyze thoroughly and objectively scientific, technical and socioeconomic information to understand the risks and impact of climatic change, as well as the possible adaptation and reduction methods of it. The IPCC consists mainly of three work groups and a special team (which makes inventories about greenhouse gases) and has published up until now four scientific reports on climate aspects, mitigation and vulnerability (the last two reports focus on climate change impact in natural and socioeconomic systems). IPCC work team I (scientific group) published in Paris, on February 2007, the fourth evaluation report about global warming. It includes the results of the studies made on the last six years, and future perspectives about their possible effects.

In Mexico's case, the scarcity of water in the north of the country is worrying, even more from the agricultural point of view. According to the results of general circulation models, a major impact due to global warming is expected in this region, and it will have important repercussions on the rural sector in the north of Mexico. This area has been subject to dry seasons for long periods of time and this situation will increase the problems of water availability for crop irrigation. The major agricultural districts are at the north of the country and they consume 80 of every 100 liters of water used in Mexico, from which 50% of water resource is wasted. Therefore, considering the potential risk and vulnerability of the area, it is important to estimate the effects of climatic change in evapotranspiration (ET) in the principal irrigation districts of the country to calculate the possible growth on water demand, for it would mean to make use of a greater water volume from the dams in a region where this resource is limited.

However, it is not easy to quantify climate change effects on agriculture for several scientific reasons; the most important of them are three: first, the estimation of the quantitative effects of global warming over future climate involves the use of numeric models on global scale that do not contain all variables influencing the phenomenon (e.g. ocean's CO_2 absorption). It also uses inconsistent data; for example the rate of greenhouse gases emission depends on the anthropogenic effect, especially in the consumption and emission of fossil fuels, choice of technology, economy evolution and demographic growth of each country. In second place the results of global models that estimate future climate (increase or decrease of temperature or precipitation) generate data that cannot be directly applied to specific regions, so it's necessary to regionalize it in a process that, if not executed, directly affects the accuracy of the calculations. In third place, empiric (temperature and radiation models) and semi-empiric (Penman-Monteith models) evapotranspiration models were not designed to estimate ETo under climatic change conditions. In its current state, this set of equations is limited to model climate change and its effects on ETo, and can be used to perform sensitivity analysis at most.

Due to the complexity of global warming effects on temperature estimations, IPCC established different scenarios of climate change that take on account expected variability of conditions instead of a single calculation of greenhouse gas emission, and its relationship with the raise of global temperature. Each exponential shows answers based on different actions to mitigate climate change impact; for example: The higher tendency (upper exponential) is constructed under a vision of total failure in the efforts to diminish the effects of global climate change. The two center exponentials are associated with the hypothesis of the proper function of carbon market, i.e., the state and private actions are effective to diminish the greenhouse gas emission which allows a moderately lesser impact of climate change. Finally, the lower exponential of the graph assumes that all involved countries sign the Kyoto Protocol to mitigate climate change impact.

2. BACKGROUND

Based in these scenarios, efforts have been made in order to estimate climatic change effect on irrigated agriculture and dryland farming in Mexico. In dryland farming case, sensibility studies were elaborated with two specific models, the CERES-MAIZE and GFDL-R30 models. The CERES-MAIZE model simulated the performance of different corn crop productions on non-irrigated areas, with varied weather, seed and soil conditions. On the other hand, the GFDL-R30 model created theoretical agro climatic maps (i.e.), maps of potential suitable regions for certain crop production. Both models were applied in the following municipalities: Atlacomulco (state of Mexico), Izúcar and Ixcamilpa (Puebla), Tuxpan and Coatepec (Veracruz) and Magdalena and La Huerta (Jalisco) (Figure

1). These models were used considering arbitrary temperature (+2˚C and +4˚C) and pre-cipitation (±10% and ±20%) increases, and even combinations of both. The CERES-MAIZE model simulation results indicate a reduction in corn crop performance in all studied regions (except Atlacomulco), which means a negative climate change effect. Moreover the GFDL-R30 model results show that the simulated regions and their agro climatic areas are unfit for crop production under potential climatic change scenarios. For irrigated agriculture, the first studies elaborated in Mexico about climate change effects in evapotranspiration of crops was made in the Mexican Institute of Water Technology (IMTA in its Spanish acronym), about a decade ago. This preliminary study consisted in the application of a FAO radiation model to estimate the effects of climate change on ET annually in Mexico. Figure 2 shows the obtained results with the radiation model, which consist of a set of ET isolines on a national scale, calculated under a climate change scenario with actual temperature conditions. Figure 3 depicts the ET annual estimates under a 3˚C increase of global temperature scenario [1].

The results of the sensitivity analysis made in this study indicate that the most vulnerable zone is the north of Mexico, wherein the increase of ET is greater in comparison to the rest of the country; up to 8% annual average for a +3˚C growth in mean annual temperature. Although the results of this study are indicative, they allow an idea of the increment of water for crop irrigation. Even without considering a climate change scenario, an important rise in irrigation water demand is expected by 2025 in México, as explained in Table 1 (this increase is estimated by 16.5% between 1995 and 2025).

Figure 1. Location where were applied the The CERES-MAIZE models: Atlacomulco (state of Mexico), Izúcar and Ixcamilpa (Puebla), Tuxpan and Coatepec (Veracruz) and Magdalena and La Huerta (Jalisco).

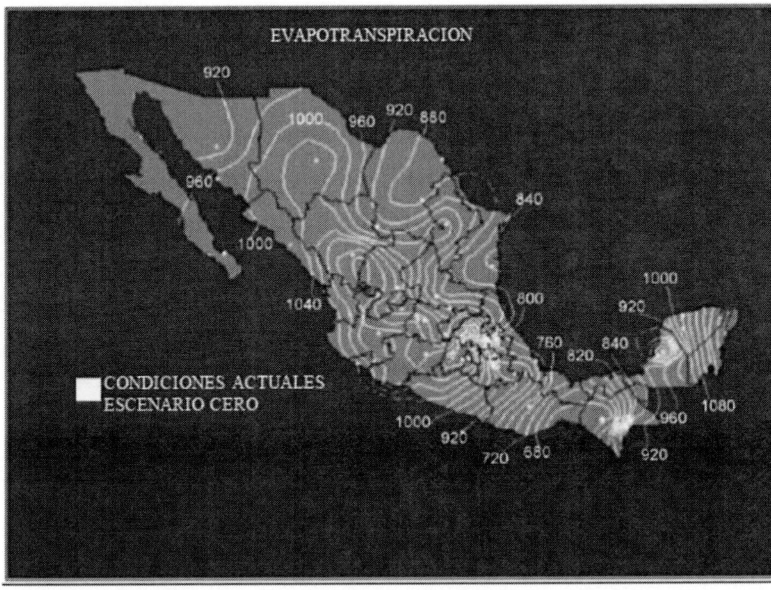

Figure 2. ET isolines under current temperature conditions [1].

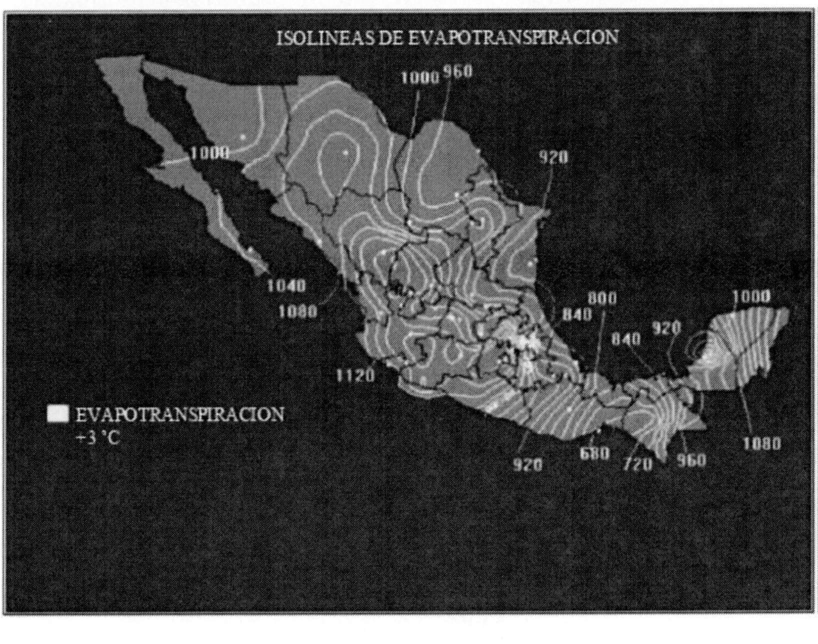

Figure 3. ET isolines with a 3°C increase on temperature [1] .

3. ET ESTIMATIONS MADE WITH HARGREAVES-SAMANI METHOD AND REGIONAL DATA

The indicative preliminary study on climate change effects in ET that used the radiation method mentioned above has several additional limitations to the ones indicated in the introduction. These are the most important: a) The IPCC scenarios were used without data on regional temperature; the process of deriving information from large scale models (global models) to regional scale models was not done; b) The same temperature increments were used for different seasons (not regionalized); c) Equal temperature increments were used for all months of the year.

A research was conducted to correct these limitations. It was made under a cooperative agreement between IMTA and the Autonomous University of Chiapas (UNACH in its Spanish acronym). The following methodology was used:

a) A method for calculating ET on the northern region of Mexico was selected. This method was to meet certain requirements, including: the usage of few variables to estimate ET, for the lack of meteorological data prevented the use of the Penman-Monteith (PM) equation; accuracy in the ET estimate results, or close to parallel the precision of the PM equation and to serve as reference for different researchers as a precise method for ET assessment.

b) The site to calibrate the method was selected. Irrigation district 041 in Valle del Yaqui, state of Sonora was chosen, due to Sonora being one of the most important states in the northern region of the country and having a network of automated meteorological stations for the calibration of the selected method.

c) The method was calibrated to estimate ET. The daily and monthly results were compared against the estimated results obtained from the witness method (Penman-Monteith ASCE).

d) The northern states of Mexico which include major irrigation districts were selected. Nine states from the north were designated, as well as 160 meteorological stations that hold records with more than 30 years worth of data.

e) Climate change scenarios were selected. Once the method for evapotranspiration estimation was calibrated, two types of scenarios were picked, the A1B and the A2. The results obtained with the A1B scenario are presented in this work.

f) Data of average anomalies in temperature obtained from IPCC models on A1B scenarios was used. The

g) results processed from 16 climate models that participated in IPCC for the fourth evaluation report were analyzed by the hydro-meteorology IMTA team using the downscaling technique. The data was later used in the research.

h) The ET data in the "zero scenario" (current conditions), and the average temperature anomalies calculated for year 2030 were compared. The ET was evaluated with the current scenario anomalies (present temperature conditions) and the estimated stipulation for year 2030; after that, a comparison was made between both sets.

Table 1. Water consumption for irrigation in Mexico, real and predicted [2].

Population ad water data				Historical data				Predictec
	1950	1960	1970	1980	1990	1995	2010	2025
Population (M)	27.49	36.71	50.4	67.05	84.5	93.97	118	136.72
Irrigation area (M ha)	1.6	3	3.58	4.98	5.6	6.2	6.8	7.1

Therefore, for a precise ET calculus under climate change scenarios (sensibility analysis) it would be convenient to use the Penman-Monteith ASCE equation for all the northern part of the country, as the one stated below [3]:

$$\lambda Et_o = \left(\frac{\Delta}{\Delta+\gamma^*}\right)(Rn-G)+\left(\frac{\gamma}{\Delta+\gamma^*}\right)\frac{k_1(0.622)\rho\lambda}{P_a}\frac{1}{r_a}\left(e_z^o-e_z\right) \tag{1}$$

where:

ET_o = Evapotranspiration of referenced crop (mmd^{-1})

Rn = Net solar radiation on the surface of the crop ($MJ \cdot m^{-2} \cdot d^{-1}$)

G = Sensitive heat flux from soil ($MJ \cdot m^{-2} \cdot d^{-1}$)

Δ = Vapor pressure curve inclination (kPa)

γ = Psychometric constant ($kPa^{\circ}C^{-1}$)

γ^* = Modified psychometric constant ($kPa^{\circ}C^{-1}$)

ρ = Density of the fluid at a constant pressure ($kg \cdot m^{-3}$)

λ = Latent vaporization heat ($MJ^{-1} \cdot {}^{\circ}C^{-1}$)

P_a = Atmospheric pressure (kPa)

e_z^o = Saturated vapor pressure of air at z height (kPa)

e_z = Vapor pressure of air at z height (kPa)

r_c = superficial resistance to vapor transfer (sm^{-1}), or simply surface resistance.

r_a = *Aerodynamic resistance to sensible heat and vapor transfer (sm^{-1}), or* aerodynamic resistance.

However this equation needs variables and data that are not available in all studied places. For this reason an analysis was made on over a dozen of empiric methods that would allow calculating with precision the ET. Thus the Hargreaves-Samani (HS) method was selected to estimates these variables. The HS method originates from [4] and [5] though various adjustments were made to the initial equation introduced on 1975. This HS equation estimates the evapotranspiration of a harvest using grass as reference. The HS equation is [3] :

$$ET_o = 0.0023 R_a (TD)^{1/2} (t_m + 17.8) \tag{2}$$

where:

TD = Difference between maximum and minimum temperature (˚C)

R_a = Extraterrestrial radiation (mmd^{-1}), (R_a Variable is obtained from tables or with a equation)

Rs = Solar radiation (mmd^{-1})

This method is recommended by several investigators, [3] [6] - [9] for its precision. It can even, if required, be used to estimate ET for periods of five days and on daily scale [10] and [11] . It is a simple method that only needs extraterrestrial radiation and minimum and maximum temperature data to calculate ET. For its calibration the daily and monthly information collected from 10 stations in irrigation district 041, Valle del Yaqui (Figure 4) was used, Stations have daily measurements of radiation, relative humidity, wind speed, temperature and precipitation; some with over ten years of data.

In Figure 5 and Figure 6 are shown the results of the comparison between the HS and the Penman-Monteith ASCE methods for the CIANO station, located in block 910, lot 3, in the following coordinates of Yaqui Valley: 27˚22'14 North latitudes, 109˚55'4 West longitudes.

Figure 4. The small circles and small rectangle represents the weather stations network, Yaqui Valley Sonora Mexico.

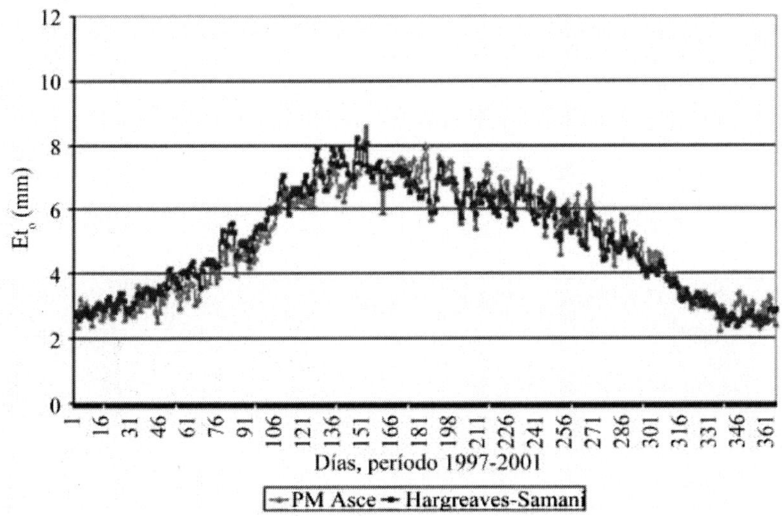

Figure 5. Penman-Monteith (PM) ASCE vs. Hargreaves-Samani comparison on a daily scale.

In both cases the linear regression coefficient is higher than 0.92, with a standard deviation from the model of 1.63 mm, a quadratic error of 0.44 in the first case and 0.6 in the second case, as well as 0.93 model efficiency for the daily scale and 0.97 for the monthly scale. Once the precision of the method was proven, the selection of the IPCC climate change scenarios proceeded. The results presented in this work were obtained under the A1B scenario considerations. The A1 family is characterized by a rapid economic growth, a global population that will reach its peak in 2025 and a society that adopts and uses efficient technologies. This family divides into two: A1F1 (intense use of petroleum fuels) and A1B (balanced use of different energy sources); the latter is considered as the most likely scenario to occur, according to the scientific community.

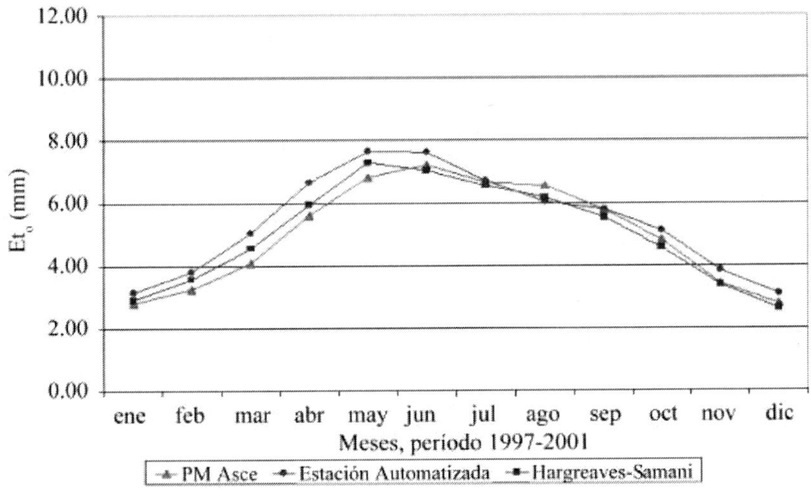

Figure 6. Penman-Monteith (PM) ASCE vs. Hargreaves-Samani comparison on a monthly scale.

When the scenario was selected, an analysis of the obtained information on the global IPCC models followed. Large scale climate models are not able to adequately simulate climate variations at regional level (regional climate behavior is a stochastic process, conditioned by global climate), therefore it is necessary to use downscaling techniques to complete the process. Downscaling techniques allow the derivation of information from global models to regional ones through the statistic inference of the relationship between both scales. Consequently, investigators from the sub-coordination of hydrometeorology of IMTA[2] studied the results of 16 climate models addressed in the Fourth Evaluation Report (2007), from where the monthly average anomalies registered during the 1960-1989 period for the A1B scenario were obtained from the models using statistic downscaling techniques. With the estimated temperature anomalies for year 2030 (in A1B scenarios), simulations were performed to calculate ET in the principal irrigation districts of the following states: Baja California, Baja California Sur, Chihuahua, Sinaloa, Sonora, Tamaulipas, Nuevo León, Coahuila y Durango (Figure 7). Figure 8 shows, in °C, the difference between the actual and the 2030 scenarios for more than 160 stations in this states (the variability observed in Figure 8 is due to regionalization of temperature through downscaling techniques).

The result of the sensitivity analysis with the HS method is shown in Figure 9, where the differences in percentage of ET between the actual and the 2030 scenarios can be observed.

Figure 10 shows the ET isolines for July in the current scenario while Figure 11 illustrates ET isolines for the 2030 scenario.

When both figures are compared, an increment by 7% in the ET is clear between the actual scenario and the 2030 scenario, as well as a major density of isolines in 2030 that indicates greater temperature gradients.

4. CONCLUSIONS

Global climate change is a theme that has gained relevance in the last decade, and one that is studied from different perspectives: socioeconomic, technical and scientific. There is worldwide concern about the potential repercussions of climate change on natural resources; in Mexico's case, the climate change effects are worrisome, especially in the agriculture sector in the north of the country for it is a wide region with scarce water resources. One of the first studies in Mexico about these potentials risks was made about a decade ago in the Mexican Institute of Water Technology. This preliminary study consisted in the application of the FAO radiation model to estimate the climate change effects on evapotranspiration, and the results obtained were a set of ET isolines under diverse scenarios that indicated the most vulnerable zone in the country, the northern region of Mexico. In this region the annual increment of ET is 8% greater than the rest of the country for a raise of 3°C in mean temperature. Due to some limitations in this preliminary investigation (e.g., global temperature data was used without regionalizing it), it was decided to make more detailed studies to estimate the climate change effects on ET on a regional scale, using the downscaling method to adjust temperature data. In this study a new methodology to estimate the ET before climate change scenarios is introduced, which includes the selection of the Hargreaves-Samani method (HS), calibrated and compared against the Penman-Monteith ASCE method in various irrigation districts in the northern part of the country, obtaining ET estimations with a 93% precision. This procedure was applied to nine states in north Mexico: Baja California, Baja California Sur, Chihuahua, Sinaloa, Sonora, Tamaulipas, Nuevo León, Coahuila y Durango. The principal results are enunciated as follows: the ET variations between the contemporary scenario and the 2030 scenario are quite significant, according to the data of 160 meteorological stations; for temperature variations between 0.1°C to 0.45°C the corresponding ET fluctuation goes from 2% in the current scenario to 7% in the 2030 scenario. These obtained percentages are greater than the ones expected to happen for the precision of the method.

Figure 7. Principal irrigationdistricts in thenorth of Mexico: Baja California, Baja California Sur, Chihuahua, Sinaloa, Sonora, Tamaulipas, Nuevo León, Coahuila y Durango.

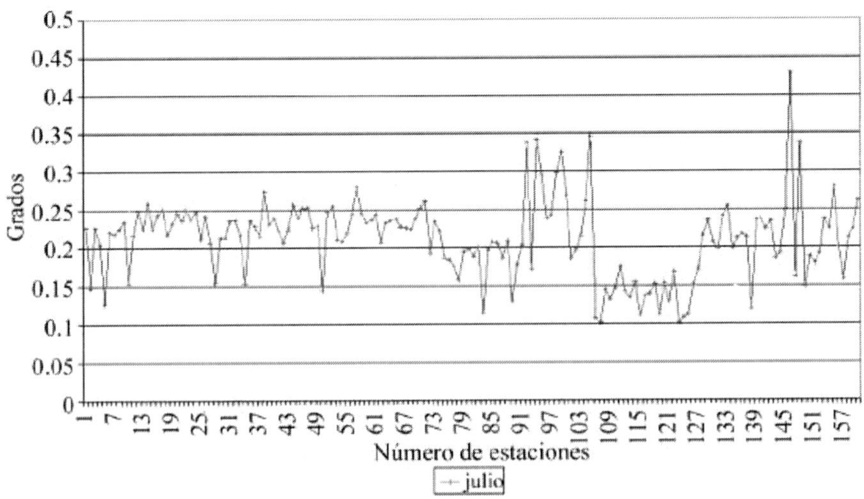

Figure 8. ET Difference between the current and the 2030 scenarios for the month of July (in °C).

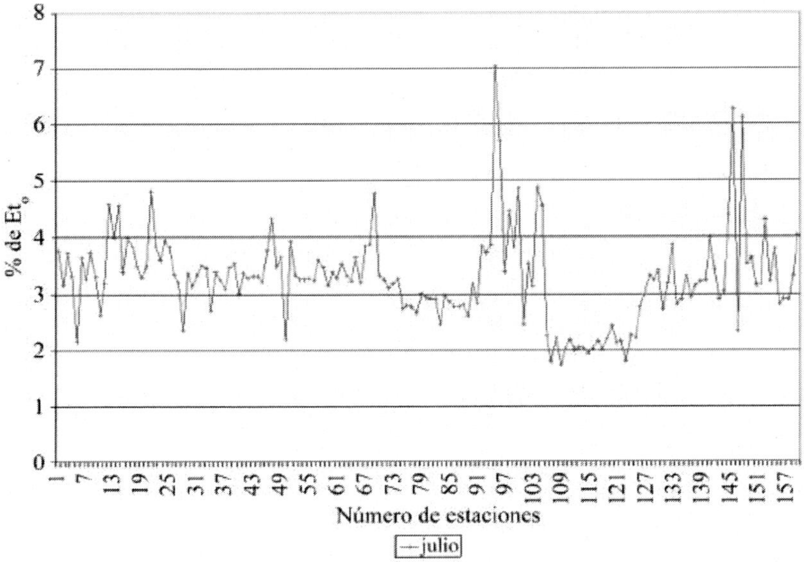

Figure 9. Percentage difference of ET for the month of July between the current and 2030 scenarios.

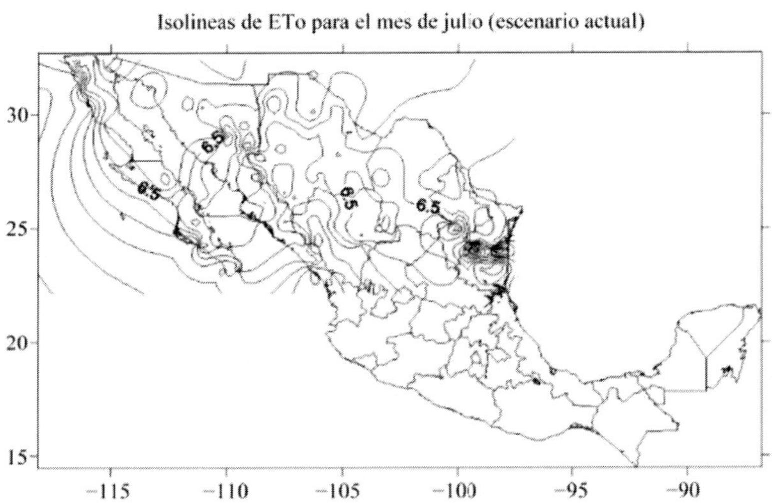

Figure 10. ET isolines in July for the current scenario.

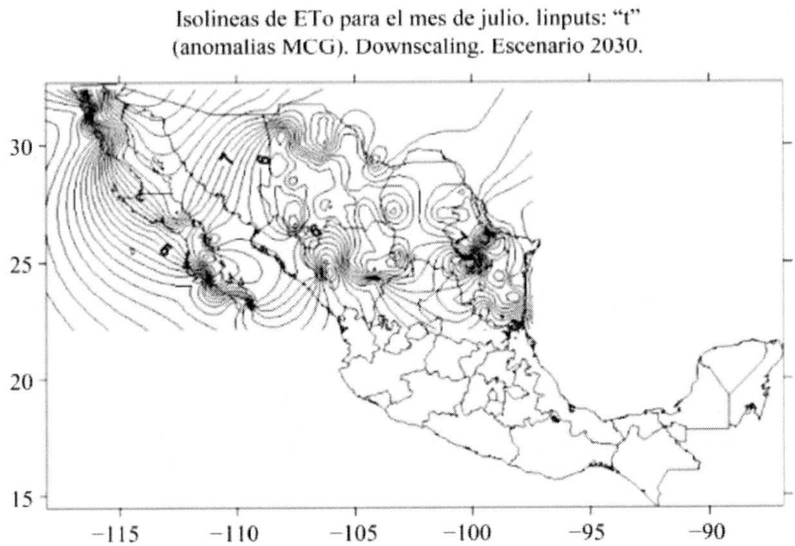

Figure 11. ET isolines in July for the 2030 scenario.

It is important to note that a 7% rise of ET (related to a regional temperature increase of approximately one degree) would represent in practice having more millions of m^3 of water in dams to satisfy the water demand of crops. With the intention to proportionate an idea of this additional water volume, data from the irrigation water volumes used in Sonora in 2004 was taken. That year, a surface of 631 200 hectares was irrigated with a total of 6315 Mm^3 of water for agricultural purposes only (superficial and underground water).

Considering that the average efficiency in water consumption for irrigation is 48%, 3283 Mm^3 of water were lost, and only 3031.20 Mm^3 were employed. The 7% of the used water volume is 211.58 Mm^3, which is almost a third of the Adolfo Ruiz Cortinez dam capacity. Therefore 211.58 Mm^3 would be the additional water demand for a 0.5°C temperature increase in Sonora. It is prudent to point out that these results are only indicative because of the limitations the evapotranspiration methods have to do ET estimations under climate change scenarios and the complexity of the theme, however, it is also important to highlight that the results allow to form an idea of the potential effect of mean temperature increase in irrigation zones. Because of the importance of the topic and the need to calculate more accurately the effects of climatic variability in irrigation and seasonal areas, the investigation must continue as well as the developing of new calculation methods that include not only the climate change effect, the needs and availability of water for crops but also the consequence of these changes on crop performance. It's important to develop and adapt methodologies that allow the estimating of crop functionality by using biological simulations, also ones that consider the potential effect of carbon dioxide and water availability. It is also important to obtain the anomaly values of "t" in a lower scale (downscaling), e.g. on a watershed scale or an irrigation district scale to obtain more accurate ET estimations under climate change scenarios.

ACKNOWLEDGEMENTS

Special thanks to Martín José Montero-Martínez and José Luis Pérez-López, IMTA investigators, for providing data to do the study of average temperature anomalies of the models, regarding mean model climatology for the 1960-1989 period (A1B scene) and Laura Montserrat Mundo-Navarro for reviewing this paper.

REFERENCES

1. Martínez, A.P., Hernández, L. and Mundo-Molina, M. (1998) Global Warming Effects on the Water Balance in México. 7th International Conference Computer in Agriculture, Florida, 26-30 October 1998. ASAE, The Society for Engineering and Agricultural, Food and Biological System.
2. Shiklomanov, I.A. (1998) World Water Resources at the Beginning of the 21st Century. Monograph Prepared and Submitted to UNESCO, Division of Water Sciences Hydrological Institute, International Hydrological Programme (IHP), UNESCO.
3. Jensen, M.E., Burman, R.D. and Allen, R.G., Eds. (1990) Evapotranspiration and Irrigation Water Requirements. ASCE Manuals and Reports on Engineering Practices No. 70, New York, 332 pp.
4. Hargreaves, G.H. and Samani, Z.A. (1982) Estimating Potential Evapotranspiration. Journal of Irrigation and Drainage Engineering—American Society of Civil Engineers (ASCE), 108, 225-230.
5. Hargreaves, G., Hargreaves, G. and Riley, J. (1985) Irrigation Water Requirements for Senegal River Basin. Journal of Irrigation and Drainage Engineering, 111, 265-275.
6. Choisnel, E., de Villete, O. and Lacroze, F. (1992) Une approche uniformisee du calcul de Lévapotranspiration potentielle pour Lénsemble des PAYS de la communaute europeenne. Unsysteme D'Information agronomique pour La Communaute Europeenne,

Centre Commun de Recherche, Commission des Communaute Europeennes.

7. Hargreaves, H.G. (1994) Defining and Using Reference Evapotranspiration. Journal of Irrigation and Drainage Engineering, ASCE, 120, 1132-1139.
8. Allen, R.G. (1995) Evaluation of Procedures for Estimating Grass Reference Evapotranspiration Using Air Temperature Data Only. Rep. Prepared for the United Nation Food and Agricultural Organization. Rome, Italy. In Jensen, D.T., Hargreaves, H.G., Temesgen, B. and Allen, R.G. (1997) Computation of Eto under Non Ideal Conditions. Journal of Irrigation and Drainage Engineering, ASCE, 123, 394-400.
9. Amatya, D.M., Skaggs, R.W. and Gregory, J.D. (1995) Comparison of Methods for Estimating REF-ET. Journal of Irrigation and Drainage Engineering, ASCE, 121, 427-435.
10. Snyder, R.L. (2000) Daily Reference Evapotranspiration (Eto) Calculator. Department of Land, Air and Water Resources, University of California. Davis, California.
11. Snyder, R.L. (2002) Epistolary Communication.

Chapter 2

Field Measurement of Cotton Seedling Evapotranspiration

R. J. Lascano[1*], J. L. Duesterhaus[2], J. D. Booker[3], T. S. Goebel[1], J. T. Baker[4]

[1]USDA-ARS[#], Wind Erosion and Water Conservation Research Unit, Plant Stress and Water Conservation Laboratory, Lubbock, TX, USA

[2]USDA-ARS, Former Postdoctoral Research Associate, Wind Erosion and Water Conservation Research Unit, Plant Stress and Water Conservation Laboratory, Lubbock, TX, USA

[3]Plant and Soil Sciences Department, Texas Tech University, Lubbock, TX, USA

[4]USDA-ARS, Wind Erosion and Water Conservation Research Unit, Plant Stress and Water Conservation Laboratory, Big Spring, TX, USA

ABSTRACT

Information on cotton evapotranspiration (ET) during the seedling growth stage and under field conditions is scarce because ET is a difficult parameter to measure. Our objective was to use weighable lysimeters to measure daily values of cotton seedling ET. We designed and built plastic weighable micro-lysimeters (ML) that were 0.35 m deep with a soil volume of 6300 cm^3. The soil core was obtained in-situ by pushing the ML well casing into the soil using a commercial soil sampler. The soil core was weighed with tension and compression type load-cells, where a change in mass of 18 $g \cdot d^{-1}$ was equivalent to a water evaporation of 1 $mm \cdot d^{-1}$. We compared load-cell measurements of changes in mass to values measured with a portable field scale by linear regression analysis, and the slope was equal to 1, indicating no statistical difference ($P = 0.05$) between the two measurements. We measured and compared seedling height, root length and leaf area of cotton plants in the ML with cotton plants in the surrounding area and this comparison showed that the ML used was suitable to measure cotton seedling ET for the first 30 days after seed emergence. The root mean squared error for crop height was 0.09 cm, for leaf area index (LAI) was 0.03 $m^2 \cdot m^{-2}$ and 6.5 cm for root length. Also, soil temperature at a 0.1 m depth was statistically ($P = 0.05$) the same in and outside the ML's. For two planting dates, we measured daily values of soil water evaporation (E) and cotton seedling ET. The day following an irrigation event, E was ~ 9 mm d^{-1} and quickly declined the following days. Results showed that ML's provide an accurate tool to measure water losses from the soil and cotton plants with a LAI of ≤0.2.

Keywords: Cotton Seedling, Lysimeter, Evapotranspiration, Transpiration, Water Use, Field Measurement

1. INTRODUCTION

Upland cotton (Gossypium hirsutum L.) is the major field crop grown on the Texas High Plains even though common early-season biological and environmental constraints can have negative effects on lint yield. For example, these negative factors include soil particle abrasion on cotton seedling plants due to high wind speed, insufficient precipitation, hail, extreme air temperatures, pathogens, and damage from pests. Of all biotic and abiotic stresses, plant available soil water is the most limiting but manageable factor for irrigated cotton production on the Texas High Plains [1] [2] . It is well recognized that the first 40 days after planting is a critical stage to manage a cotton crop to achieve maximum lint yield potential [1] -[4] . Even as irrigation water from the Ogallala aquifer declines, cotton remains a viable crop for the Texas High Plains producers [5] . A great deal of knowledge has been gathered on cotton physiology, water and nutrient requirements, and other environmental factors that affect cotton productivity; however, much of this work has focused on the middle to late vegetative and late reproductive stages of cotton growth stages. Information on cotton evapotranspiration (ET) under field conditions and in the early stages of crop growth is limited.

The developmental stages of cotton are divided into five growth stages: 1) germination and emergence; 2) seedling establishment; 3) leaf area and canopy development; 4) flowering and boll development; and 5) maturation [4] [6] . Also, the cotton plant, given its indeterminate growth habit, i.e., simultaneously producing vegetative and reproductive growth, for a long period of time (130 - 180 days) makes cotton structurally complex compared to other field crops [7] . The wide range of soils and climates where cotton is grown along with its indeterminate growth habit is a challenge to manage the many biotic and abiotic stresses that impact cotton lint yield [5] [8] .

The effect of biotic and abiotic stresses on cotton seedling emergence, growth and development is the subject of several studies. For example, soil water content a few days after soil abrasion was key to plant survival, and 3-day-old cotton plants had a higher survival rate with increased soil water compared to 9-day-old plants with the same soil water content [9] . Krieg and Carroll [10] reported that minimal air temperatures limited non-lipid metabolism causing reductions in cotton seedling growth rates. McMichael and Quisenberry [11] discussed several soil environmental factors including temperature, soil water and nutrients, and soil strength that impact cotton root growth and development. Reddy et al. [12] found that cold air temperatures during seed emergence reduced cotton seedling biomass and subsequent growth rates. Pace et al. [13] found that while cotton root dry weight was less in the non-drought stressed plants, tap-root length was greater in the non-stressed plants. Baker [14] studied the abrasion of cotton seedlings and their recovery from windblown sand. The above examples illustrate the impact of biotic and abiotic stresses on cotton seedlings and their subsequent effects on ET and lint yield.

Measuring crop ET under field conditions is challenging even on well-developed crops, and is more complicated during the early stages of plant growth when the leaf canopy is small and the leaf area index (LAI) is <1. Under field conditions, a common method to measure crop ET and its two components, soil water evaporation (E) and plant transpiration (T), is gravimetrically by using weighing lysimeters. A discussion on the history and design of lysimeters and their use for measuring water use by vegetation is given by [15] and a review of methods to measure and calculate crop ET is given by [16] . In general, weighable lysimeters are useful to measure crop ET, but often they are large and expensive to install and cannot partition E and T. Examples of large weighing lysimeters are given by [15] [17] -[21] . Smaller lysimeters, i.e., mini- and micro-lysimeters, have also been used to measure E, T and ET on smaller scales and examples are given by [22] -[31] . Hereafter, no distinction is made between mini- and micro-lysimeters and they are abbreviated as ML.

The ML's have been coupled with other instrumentation, including dual-probe heat-pulse, stem flow gauges, and time domain reflectometery to measure the amount of soil water loss through E and ET [31] -[35] . These ML's are normally installed and weighed early in the morning and then again weighed in evening hours using a portable scale and thus provide an estimate of day time values of soil E and ET. In addition, load cells have been used with ML's providing an automated and continuous record of weight, which can then be used to calculate evaporative losses of water due to E and ET [28] [36] [37] .

The concept of designing a ML to measure E, T and ET requires isolating a soil volume that can repeatedly be weighed and then use changes in mass to calculate the rate of water loss. It is assumed that inside and outside soil conditions are similar and thus it is of particular importance to minimize the disturbance of the soil surface in the ML [23] . With time the soil water content and temperature profile inside the ML will deviate from that of the surrounding soil and after several days E and T from the ML may no longer be representative of the surrounding field, particularly when plants are present. However, these limitations are minimized by using a ML material with low thermal conductivity such as plastic and by using lengths ³ 0.3 m as suggested by [26]. In general and for a bare soil, the wall material of the ML affects the vertical temperature distribution, but not E, provided that ML's are replaced every day [25] [27] . When a plant is in the ML, root growth will be constrained by the inside soil volume and after several days the measured ET may be underestimated, compared to the ET of the surrounding field. The larger the soil volume of a ML the longer a single ML can be used to measure ET before it has to be replaced and the measured ET values are not representative. It is evident that ML is a tool that has many advantages measuring cotton ET of plants under field conditions and in the initial stages of crop growth stage, i.e., seedling stage.

Other methods available to measure crop ET under field conditions lack the necessary resolution to measure the daily ET during the early stages of crop growth [16] . For example, by measuring the soil volumetric water content (q_v) over the depth of the root zone it is possible to calculate the change in soil water storage and estimate crop ET. These measurements of q_v require calibrated soil water sensors that respond to electromagnetic properties, such as time domain reflectometery, ground penetrating radar, and capacitance methods [38] -[40] . The well-known neutron attenuation method is not included because of its limitation to correctly measure the q_v in the surface layer, where fast neutrons escape to the atmosphere making it difficult to accurately measure the soil water content of the surface layer. The hourly and daily crop ET of cotton plants during the seedling stage is on the order of 0.5 $mm \cdot h^{-1}$ and 3 $mm \cdot d^{-1}$, respectively [41] . For example, a daily T of 3 mm for a 0.3 m root-depth represents a change in q_v of ± 0.01 $m^3 \cdot m^{-3}$, which exceeds the accuracy of the majority of the commercial sensors available to measure q_v [42] . Therefore, the measurement of daily values of crop ET in the early stages of crop growth and under field conditions require a gravimetric method such as lysimetry or by using whole plant chambers such as those described by [43] [44] . In these chambers, differentials of incoming and outgoing H_2O and CO_2 concentrations are used to calculate plant ET and net assimilation using an infrared gas analyzer. These chambers provide an excellent tool to measure whole plant ET from emergence to first-square; however, the necessary equipment is expensive. Conversely, gravimetric measurements simply require a weighing scale and in terms of equipment are less expensive than other methods.

On the Texas High Plains, plant available soil water is the most limiting factor in cotton production and the first 40 - 60 days after planting is a critical stage to manage the cotton crop to achieve maximum lint yield potential. However, information on water use by cotton plants from emergence to first-square, in the seedling growth stage, and under field conditions is rare. Therefore, the objective of this study was to evaluate the measurement of seedling cotton ET using lysimetric techniques under field conditions.

2. MATERIALS AND METHODS

The laboratory and fieldwork were done at the facilities of the United States Department of Agriculture-Agri- cultural Research Service (USDA-ARS), Cropping Systems Research Laboratory in Lubbock, TX (33.59°N, 101.89°W, and average elevation of 960 m above sea level). Laboratory evaluations were done during the fall season of 2009 and field measurements were done during the summer of 2010. In our evaluation of ML's to measure cotton seedling ET we built ML's with and without load cells. Our field evaluations were done with cotton planted on two dates, i.e., 12 May and 22 June 2010. The layout of the experimental field used to evaluate the ML's is shown in Figure 1(a) and the location where ML's were installed is shown in Figure 1(b).

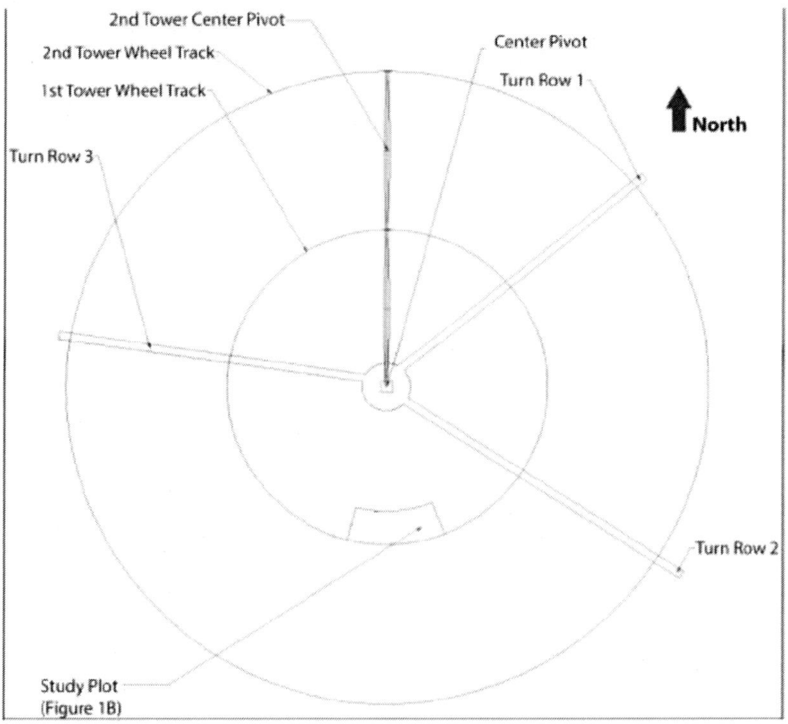

(a)

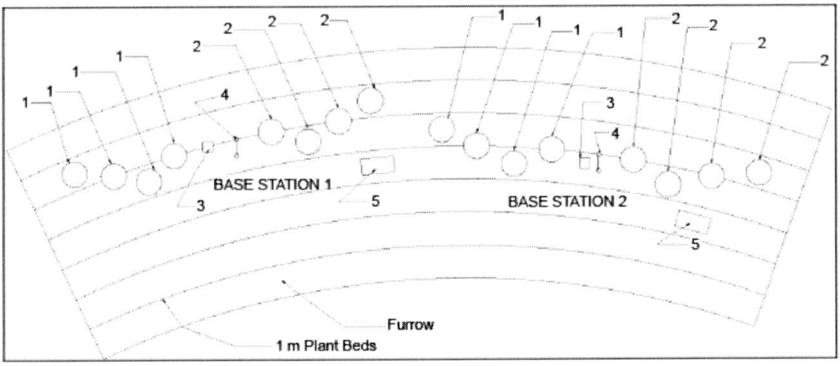

(b)

Figure 1. (a) Layout of the experimental field used to evaluate micro-lysimeters, showing the 110 m long center pivot with two spans and turn rows used to access the study plot. The center pivot is located on the facilities of the Cropping Systems Research Laboratory, USDA-ARS in Lubbock, TX (33.59°N, 101.89°W). (b) The experimental plot, 8 rows wide and 25 m long, showing location of micro-lysimeters (ML) to measure soil water evaporation (E) and cotton seedling evapotranspiration (ET). The numbers 1 and 2 are locations where ML's were installed in the planted row or in between rows, 1 m apart. The Base Station is the location of portable field scale. Number 3 shows location of rain gauges; number 4 shows location of soil thermocouples, and 5 shows location of datalogger used to weigh ML's with load-cells. The field plot was ~200 m^2.

It was of interest to determine the duration whereby the ML's used to measure the seedling ET were representative of cotton plants in the surrounding area and thus we measured and compared the crop height, leaf area and root length (RL) of cotton seedlings in the ML's to cotton plants in the surrounding area. These measurements were done on cotton seedlings for both planting dates. Crop height was measured every day after emergence on cotton seedling in the ML's (Figure 1(b)) and on ten cotton seedlings in closed proximity. At the end of each of the planting trials, we measured in the laboratory the leaf area of cotton seedlings in each ML's using a leaf area meter (Model LI-3100C Area Meter, Li-Cor, Lincoln, NE). The leaf area of cotton seedlings on the surrounding area was measured on plants harvested in three 1-m lengths of row. These plants were collected on the same date as those plants from the ML's. The RL was measured using the line-transect method of Newman [45] on the cotton roots of each plant in the ML that were washed with water and separated from the soil. The leaf area and RL was measured in each of the 8 ML's shown in Figure 1(b).

Soil temperature was measured inside and outside ML's using Type-T thermocouples connected to a datalogger (CR3000 Campbell Sci. Inc., Logan,

UT) from the 7 to the 21 June and from the 1 to 31 July in 2010. Thermocouples were installed at a depth of 10 cm from the soil surface inside the ML and directly into the soil in close proximity to the ML. These measurements were replicated at four locations in the field (Figure 1(b)).

2.1. Site Description

The soil of the experimental area is classified in the Amarillo soil series (fine-loamy, mixed, Thermic, Aridic Paleustalf) and soil physical and hydraulic properties are given by [46] . This area is characterized by a semi-arid climate, with warm summer air temperatures, high wind speed and erratic rainfall distribution. For Lubbock, TX the long-term monthly average grass reference evapotranspiration (ET_o) ranges from a low of 2 $mm \cdot d^{-1}$ in December to a high of 7.5 $mm \cdot d^{-1}$ in June. In Lubbock, TX the mean annual rainfall is 474 ± 144 mm and the distribution is characterized by a relatively dry winter period with most of the rain during the growing season, May to September [8] [47] . The experimental site was irrigated with a 110-m long center pivot (Zimmatic[1], Omaha, NE) system with two spans, covering ~4 ha (Figure 1(a)).

Experimental Plot. The outside span of the center pivot was planted with grain sorghum (Sorghum bicolor) and the inside span was planted with cotton. All planting was in circular bedded rows that were ~1 m apart. An area, 200 m^2 (8 rows of cotton and ~32 m long) within the inside span, 32 m from the center of the pivot, was used to evaluate the ML's to measure cotton ET (Figure 1(b)). To evaluate the ML's under field conditions, cotton (FiberMax-9180Ò, Bayer Crop Science, Research Triangle Park, NC) was planted on two dates on 1-m bedded rows as shown in Figure 1(b). The planting date for the first field trial was 12 May 2010 and 22 June 2010 for the second planting date. Emergence dates were 23 May and 27 June for each planting date, respectively. Stand counts averaged 20 $plants \cdot m^{-1}$ for both planting dates. The purpose of these two planting dates was to have access to cotton plants in the seedling growth stage and evaluate the ML's to measure E from bare soil and cotton seedling ET. Grain sorghum (NK7633, Chromatin, New Deal, TX) was planted, ~1 m bedded circular rows, on the 4 May 2010 in the outside span of the center pivot (Figure 1(a)).

2.2. Micro-Lysimeters (ML)

The weighable ML's designed and used in our experiments consisted of four basic components and the item number given in the ML description are shown in Figure 2. A lysimeter that holds the soil and/or plant (item 4); a rubber cap at the bottom of the lysimeter (item 5); load-cell (item 2) that weighs the lysimeter and two plates (item 3); and casing that allows the lysimeter to freely move and be weighed with the load-cell (item 6). The lysimeter was constructed from 15.2 cm inside diameter schedule 40 polyvinylchloride (PVC) pipe cut to 35 cm lengths (item 4). This size was selected because the plant available water for an Amarillo soil is about 0.25 $m^3 \cdot m^{-3}$ [46] and thus the soil in the ML could store ~90 mm of water, which could supply cotton seedlings ~3 $mm \cdot d^{-1}$ of water for 30 days. We selected and used PVC pipe because of its low thermal conductivity (0.19 W $m^{-1} \cdot °C^{-1}$) and because lengths ^{3} 0.3 m allow measurements up to nine consecutive days without having to change the ML's as suggested by [26] .

A critical element in the design of a weighing ML is that the lysimeter (item 4) holding the soil and/or plant should be able to move freely without obstruction from another wall or part of the lysimeter. For this purpose, and to measure the changes in mass as a function of time we built sleeves, i.e., casing walls (item 6) constructed of 20.3 cm inside diameter schedule 40 PVC pipe cut to 56-cm length. The casing ensured that the soil around the ML would not collapse and allowed the installation of a load-cell (item 2) to weigh the ML. Coarse sand was used to form a 10 cm base at the bottom of each casing well to allow water drainage (item 7). When the ML's were positioned in the casing wells, the top of the ML was ~1 cm above the surface of the soil.

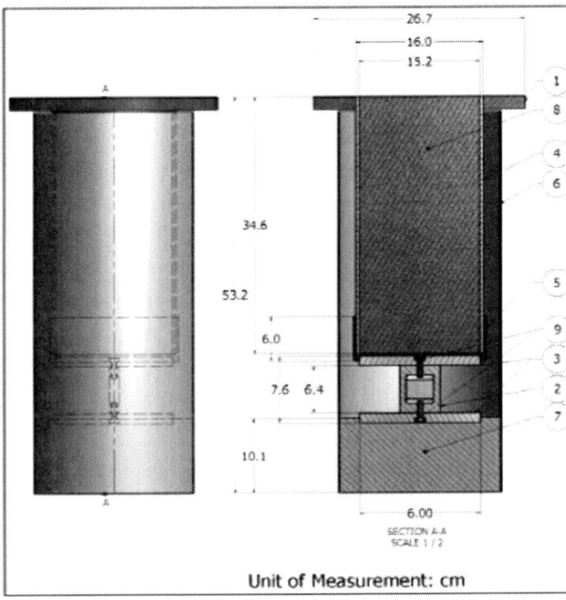

PARTS LIST

ITEM	QTY	PART NUMBER	DESCRIPTION
1	1	Heat Shield	R-Tech Styrofoam Insulation
2	1	Load Cell	"S" Type Compression Load Cell
3	2	Load Cell Plate	6" Dia x .5" Aluminum Plate
4	1	Lysimeter	6" PVC pipe with cap - holds soil and plant material
5	1	Lysimeter Cap	Rubber cap to hold soil in Lysimeter
6	1	Lysimeter Well Casing	8" PVC pipe - shields Lysimeter from surrounding soil
7	1	Sand	Large grained soil - allows for water drainage in Microlysimeter
8	1	Soil	Media in which plant is grown
9	2	ANSI/ASME B18.2.1 - 1/4-20 UNC - 0.875	Hex Bolt - UNC (Regular Thread - Inch)

Figure 2. Design of micro-lysimeters used to measure soil water evaporation (E) and cotton seedling evapotranspiration (ET). Also given are parts list with item number (1 - 9), quantity of each item, part number and description.

Load-Cells. To measure changes in mass with the ML's we selected s-beam tension and compression type load-cells (LSB302, Futek Advanced Sensor Technology Inc., Irvine, CA) shown as item 2. These load-cells have a maximum weighing capacity of 23 kg. The load-cell assembly was comprised of two circular aluminum plates (12.7 mm Height ´ 152.4 mm Diameter) (item 3) that were bolted (item 9) to each load-cell (item 2). The load-cell electromotive force (emf, mV) output was sampled with a datalogger (CR3000 Campbell Sci. Inc., Logan, UT) at a frequency of 0.2 Hz and averages were calculated and stored at 30 min intervals. The wires, 3 - 6 m in length, of each load-cell were taped to the inside wall casing of the lysimeter emerging at the top and then connected to the datalogger. Care was taken that the load-cell wires did not interfere with the free movement of the lysimeter.

Load-Cell Calibration. Each load-cell used in the ML's was calibrated under laboratory conditions by placing certified calibration weights (Mettler-Toledo, Columbus, OH) on the load-cell plate, over a 0.01 to 20.0 kg range, and registering the corresponding emf output (mV). For the ML's built and selected load-cells a change in mass of 18.15 $g{\cdot}d^{-1}$ was equivalent to an evaporative water loss of 1.0 $mm{\cdot}d^{-1}$.

Load-Cell Evaluation. Daily measurements of E and ET obtained with ML's with load cells were compared to values measured with a portable field scale (Model CBC32, Adam Equipment Co., Danbury, CT, maximum capacity of 32 kg, readability of 1 g). For this purpose, sixteen ML's divided into two groups of eight, were selected and installed as shown in Figure 1(b). In this arrangement, 8 ML's were on either side of the bed, giving a measure of E and the other eight ML's were on the planted row, giving a measure of cotton seedling ET. All 16 of these ML's were manually weighed daily at 0700 h CST for a period of 15 days during the first planting period and 18 days during the second planting period.

Field ML Installation. An undisturbed in-situ soil core was obtained by setting the lysimeter (item 4) on the soil surface and placing an aluminum plate (0.2 ´ 0.3 m) on top that was used to push item 4 into the soil using a soil-sampling machine (Giddings Machine Company Inc., Windsor, CO). In this step care was taken to minimize compacting the soil inside the lysimeter and disturbing the soil surface. The lysimeter was pushed until the top was flush with the soil surface and thereafter excavated with shovels and the 'soil core' was carefully cut at the bottom and rubber cap (item 5) was then placed on the bottom. The 'soil core' with the rubber cap on the bottom was moved over and placed in a casing (either on top of a load cell or on an additional 10 cm of sand in place of a load cell). The dug hole was made deep and large enough to accommodate item 6, such that the top was ~1 cm above the soil surface. With item 6 in place, the bottom was filled with a 10 cm layer of coarse sand (item 7), and the bottom load-cell plate (item 3) along with the load-cell (item 2) and top load-cell plate were assembled using two bolts (item 9). The final step in the field installation was to place a circular piece (item 1) of 20 mm thick Styrofoam insulation (R-Tech®, Insulfoam, Tacoma, WA) around the outer edge of the lysimeter as shown in Figure 2. With the heat shield in place, item 4 was placed on top of the load-cell plate.

The installation of ML's without load-cells was similar as previously described, except that only the in-situ soil core (item 4) was interred. Again, item 4 was inserted by placing the aluminum plate on top of the well casing and pushing into the soil using the Giddings soil-sampling machine. Once the top of item 4 was flush with the soil surface a rubber cap (item 5) was placed on the bottom of the excavated soil core. Again, care was taken to minimize disturbance of the soil surface and compacting the soil inside the lysimeter.

3. RESULTS AND DISCUSSION

Our objective was to evaluate weighing ML's to measure the daily ET of cotton seedlings under field conditions. This objective was prompted by the lack of information on the water use of cotton in the early stages of growth, which is critical to manage and achieve maximum lint yield potential. We selected a gravimetric method that when used to measure ET requires isolating a soil volume with plants growing in it that can be weighed repeatedly. The plants growing in the lysimeter must be similar to surrounding plants so that the in-situ measurements of crop ET are representative of field conditions.

3.1. Leaf Area Index (LAI) and Root Length (RL)

Using plant height (data not shown) as a criterion to determine when cotton seedlings in the ML were different than plants in the field we determined that the ML's used provided a period of about 30 days after emergence (Table 1). The root mean square error (RMSE) for the crop height was 0.9 cm. The first planting date (12 May) and the second planting date (22 June) provided 15 and 18 days, respectively of comparisons of load-cell and manually weighed measurements of cotton seedling ET. However, we also compared the measured LAI and RL of cotton seedlings in and outside the ML, and these results are given in Table 1. For both planting dates, the average measured LAI of cotton seedlings inside and outside the ML was similar. During the first planting date the average LAI of cotton seedlings in the ML, increased from 0.08 $m^2 \cdot m^{-2}$ to 0.25 $m^2 \cdot m^{-2}$ over a period of 13 days and the corresponding LAI, outside the ML's, increased from 0.07 $m^2 \cdot m^{-2}$ to 0.28 $m^2 \cdot m^{-2}$. For the second planting date, the average LAI of plants in the ML's increased from 0.04 $m^2 \cdot m^{-2}$ to 0.13 $m^2 \cdot m^{-2}$ over a period of 22 days and for the field grown plants the average LAI increased from 0.03 $m^2 \cdot m^{-2}$ to 0.18 $m^2 \cdot m^{-2}$. The RMSE for LAI was 0.03 $m^2 \cdot m^{-2}$.

Table 1. Average measured leaf area index (LAI, $m^2 \cdot m^{-2}$) and root length (RL, cm) in the micro-lysimeter (ML) and field and corresponding standard error (SE) for the first and second planting dates. Also, given is the day of year (DOY) and days after seed emergence. The ML measurements were done on the eight cotton plants shown in Figure 1(b).

Date, 2010	DOY	Days after Emergence	Average ML LAI ± SD [$m^2 \cdot m^{-2}$]	Average Field LAI ± SE	Average ML RL ± SE [cm]	Average Field RL ± SE
			First Planting Date—12 May			
9 June	161	17	0.08 ± 0.011	0.07 ± 0.020	35.1 ± 2.5	41.2 ± 2.2
14 June	166	22	0.14 ± 0.018	0.12 ± 0.022	43.9 ± 2.2	46.6 ± 3.2
22 June	174	30	0.24 ± 0.024	0.28 ± 0.030	45.3 ± 2.5	52.2 ± 3.2
			Second Planting Date—22 June			
7 July	189	6	0.04 ± 0.001	0.03 ± 0.001	19.9 ± 2.3	21.5 ± 2.1
14 July	196	13	0.07 ± 0.005	0.09 ± 0.024	37.1 ± 2.3	39.9 ± 2.9
22 July	204	21	0.11 ± 0.007	0.12 ± 0.022	52.1 ± 2.9	55.2 ± 2.1
29 July	211	28	0.13 ± 0.007	0.18 ± 0.017	51.7 ± 1.3	65.2 ± 3.5

For both planting dates, the measured RL of cotton seedlings in the ML were slightly lower than for plants grown in the surrounding area (Table 1). For example, for the first planting date the average RL for cotton seedlings in the ML's increased from 35.1 cm to 45.3 cm over a period of 13 days and the corresponding RL for outside plants increased from 41.2 cm to 52.2 cm. A similar result was obtained for the second planting date. However, for both planting dates, the coefficient of variation (standard deviation/average) of measured RL was on average 18%, indicating that statistically there was no difference of RL for cotton seedlings in and outside the ML. The RMSE for the RL was 6.5 cm.

3.2. Soil Temperature

An example of hourly measured soil temperature on a single ML for a 2-day period, 12 - 13 June (DOY 164 and 165), and in the surrounding soil is shown in Figure 3. During early morning hours the ML soil temperature is ~0.5°C cooler than the surrounding soil and as the day progresses the ML soil temperature increases and is ~1°C warmer at 1800 hours. The two soil temperatures tend to equilibrate, as expected, during nighttime, and the pattern from the previous day repeats (Figure 3). The average hourly soil temperature in and outside the 4 ML's for a 12-day period from 11 to 21 June (DOY 161 to 173) is shown in Figure 4 and the linear regression comparing the inside and outside soil temperature is shown in Figure 5. The hourly pattern of measured soil temperature in and outside a single ML (Figure 3) is repeated for the 12 days shown in Figure 4. The soil temperature in the ML is cooler that the outside soil in the morning hours before sunrise and as the day progresses the ML soil temperature is warmer than the outside soil (Figure 4). The linear regression shows a correlation coefficient (r^2) of 0.91 and a slope of 0.76 over a range from 25°C to 33°C (Figure 5). These results show that the in-situ field installation of ML's had a minimal impact on the soil temperature inside and outside at a depth of 0.1 m across the temperature ranges reported in this study. This comparison was done for a continuous 12 day (Figure 4) and 33 day (DOY 176 - 209) period (data not shown). For example, on DOY 209, after 33 days of the ML remaining in-situ, the hourly pattern shown on DOY 164 - 165 (Figure 3) was the same. From these measurements we can conclude that the effect of isolating a soil volume with a ML on soil temperature at a 0.1 m depth for a 30-day period is minimal.

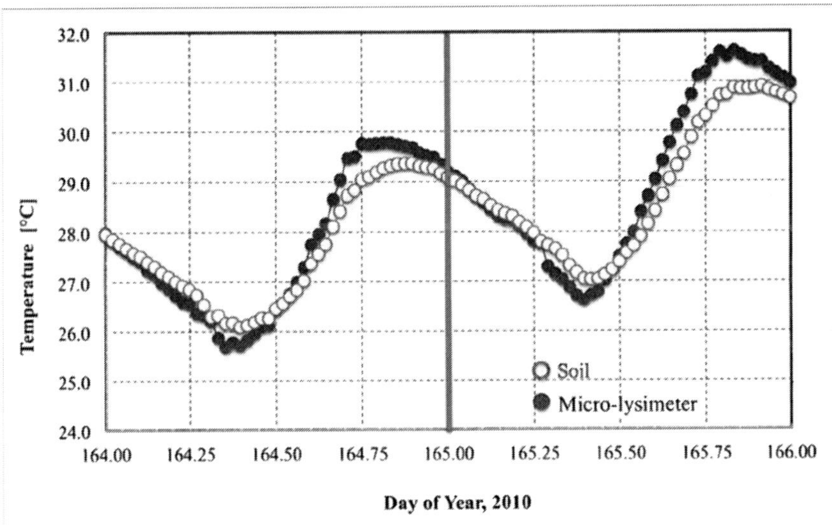

Figure 3. Measured soil temperature, 10 cm depth, inside the micro-lysimeter (ML) and in the soil in close proximity to the ML.

3.3. Soil Water Evaporation (E) and Cotton Seedling Evapotranspiration (ET)

The measurement of hourly and daily soil water E and plant ET with in-situ ML's under field conditions is a gravimetric method that is accurate and involves the repeated measurement of the mass of an isolated soil volume as a function of time [16] . To obtain hourly values of cotton seedling ET the ML's were weighed with load-cells (Figure 2) and daily values of ET were compared to values obtained with a portable field scale. The comparison of daily values of E and ET obtained with the portable field scale and load-cells are shown in Figure 6. The linear regression for the cotton seedling ET was y = 1.01x, where y is the scale ET in $mm \cdot d^{-1}$ and x is the corresponding load-cell ET value in $mm \cdot d^{-1}$. For E, the linear regression was y = 0.99x, where y is the scale-E in $mm \cdot d^{-1}$ and x is the corresponding load-cell value of E in mm·d^{-1}. We tested if the slopes (P = 0.05) where different than 1, i.e., dy/dx ¹ 1, showing that there were no differences between values of E or ET measured with the load-cells or portable field scale. Both liner regressions were forced through the origin and yielded an $r^2 = 0.87$ for the ET measurement and an $r^2 = 0.99$ for the E measurements. These results confirm that a gravimetric method to measure E or ET is accurate [16].

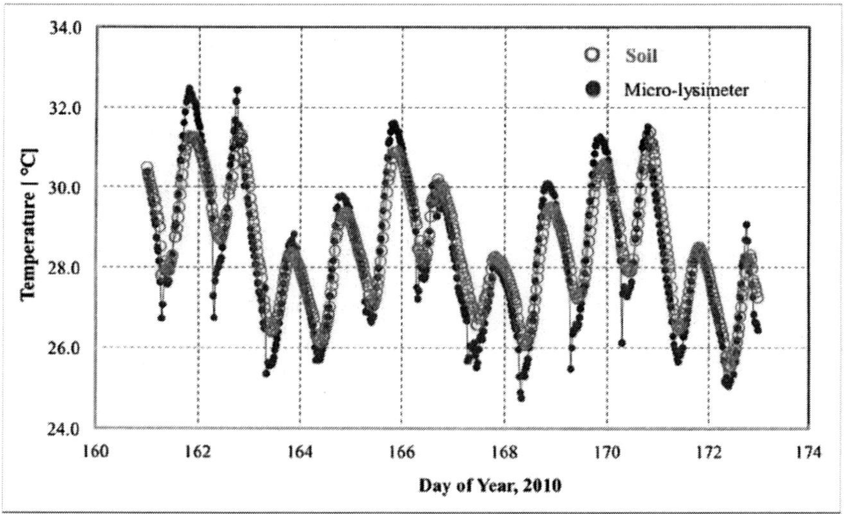

Figure 4. Measured average soil temperature inside and outside four micro-lysimeters for a period of 12 days from DOY 161 to DOY 173 in the experimental plot shown in Figure 1(b).

Examples of the 1/2-hourly record of the mass of a single ML on bare soil, a measure of E, and of a ML with a cotton seedling, a measure of ET, are shown, respectively in Figure 7 and Figure 8. The measurement of 1/2 hourly values of E was for a 12-day period from DOY 161 to 173 (Figure 7). At 0850 hours on DOY 161 the initial mass was 12.5 kg and remained relatively constant until 1800 hours, increasing to 12.9 kg. This increase in ML mass was a result of a 25-mm irrigation applied by the overhead sprinkler (Figure 1(a)). Thereafter, the mass of the ML gradually decreases as a result of water evaporation from the bare soil. The increase in mass on DOY 170.73 (1700 hours) from 12.3 kg to 12.6 kg was again due to a 20-mm irrigation. The example for the 1/2 hourly values of ML mass with a cotton seedling, also show the increase in ML mass due irrigation applied by the center pivot at the end of DOY 161 and on DOY 170.73 (Figure 8).

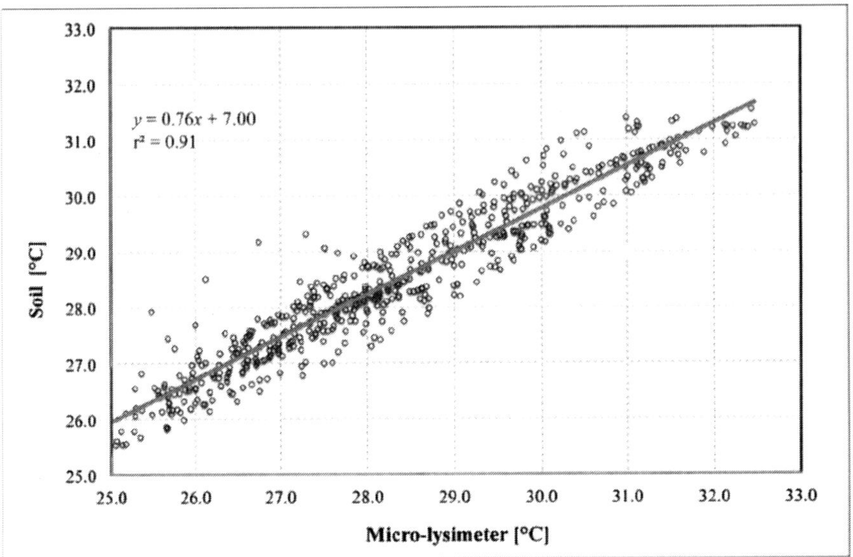

Figure 5. Linear regression and coefficient of correlation (r^2) of measured soil temperature inside and outside micro-lysimeters for the 12-day period from DOY 161 to DOY 173.

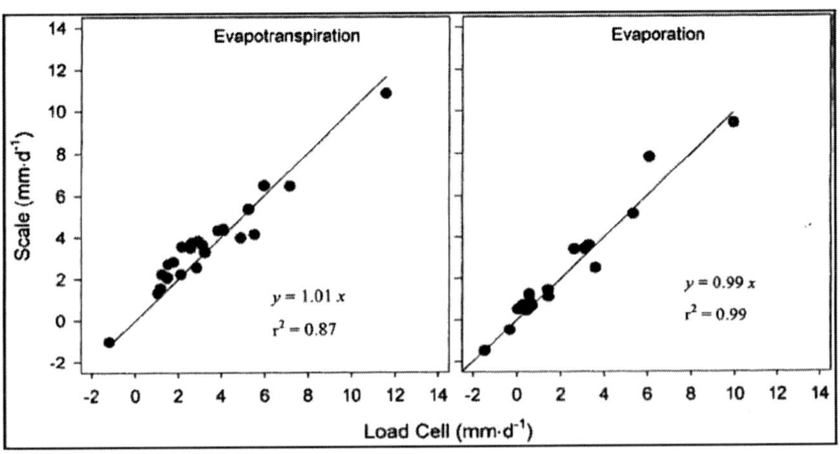

Figure 6. Linear regression and coefficient of correlation (r^2) for daily values of evapotranspiration ($mm{\cdot}d^{-1}$) and soil water evaporation ($mm{\cdot}d^{-1}$) measured with the portable field scale and micro-lysimeter load-cell.

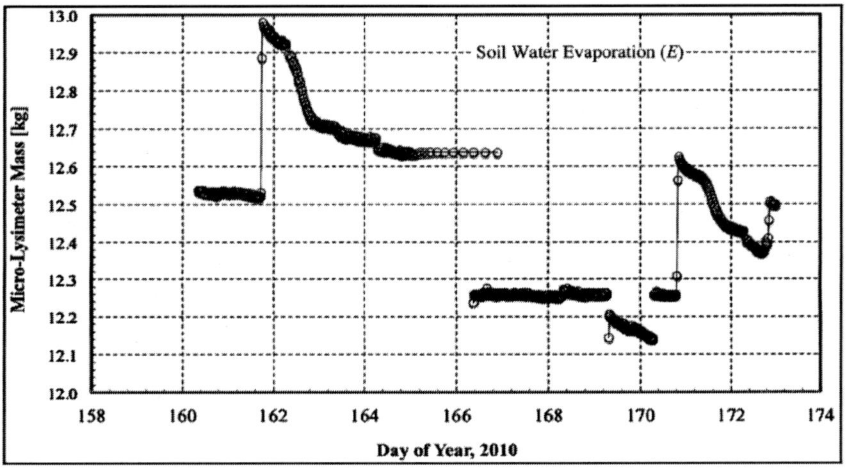

Figure 7. Half-hourly micro-lysimeter mass (kg) measured with a load-cell on a bare soil for 12 days from DOY 161 to DOY 173 and used to calculate soil water evaporation (E) from bare soil.

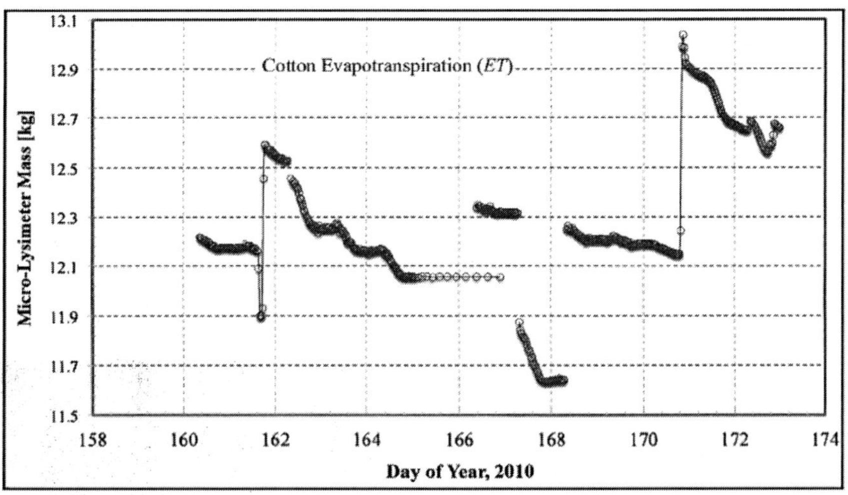

Figure 8. Half-hourly micro-lysimeter mass (kg) measured with a load-cell on a cotton seedl- ing for 12 days from DOY 161 to DOY 173 and used to calculate cotton seedling evapotranspiration (ET).

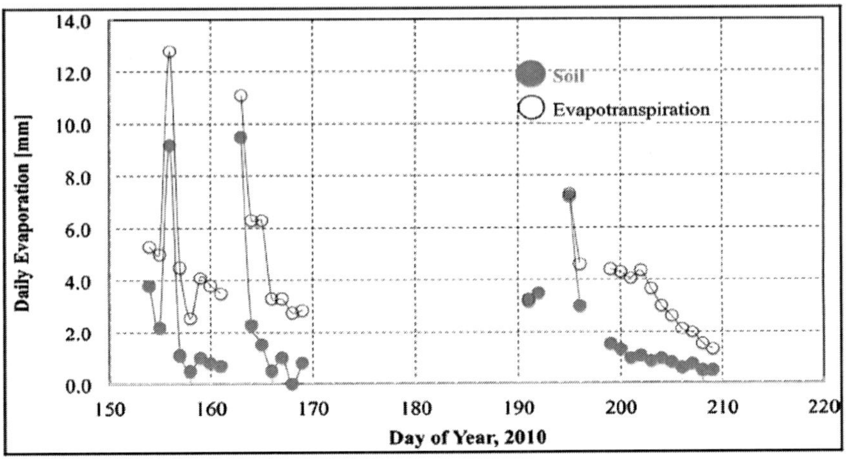

Figure 9. Average daily values of measured soil water evaporation (E) and cotton seedling (ET) for both planting dates. The first planting study started on DOY 154 and ended on DOY 169. The second planting study started on DOY 191 and ended on DOY 209.

The average daily values of E and ET measured with the weighable ML's on both planting dates are shown in Figure 9. The gaps in the data were due to rainfall that precluded access to the plot to weigh ML's with the portable field scale. The values of E ranged from a low of ~0.5 mm·d^{-1} on several days (DOY 158, 166 - 167, and DOY 208 - 209) to a high of 9.2 mm·d^{-1} on DOY 156 and 9.5 mm·d^{-1} on DOY 163 for the first planting date and a high of 7.2 mm·d^{-1} on DOY 195 for the second planting date. These high values of E can be expected on the Amarillo soil series, particularly after an irrigation or rainfall event where a high rate of E is followed by a much lower value immediately afterwards, i.e., 9.5 mm·d^{-1} on DOY 163 to 2.3 mm·d^{-1} on DOY 164, and 1.5 mm on DOY 165 [48] [49] . The daily values of ET ranged from a low of ~0.5 mm·d^{-1}, measured on several days on both planting dates, to high values of 12.8 mm·d^{-1} on DOY 156 and 11.1 mm·d^{-1} on DOY 163. However, the majority of ET was due to a high value of E. The ability to separate E from ET provides an estimate of plant transpiration (T) values, which are low early in the growing season and increase as plants grow and develop [41] [49] .

Information on evaporative losses of water from the soil and the plant, throughout the growing season, is an important subject in cropping systems research for both dryland and irrigated crops [41] [50] . In managing water from either rain or irrigation, a frequent goal is to minimize the evaporative losses of water from the soil and to increase crop transpiration. This requires information on two of three components of ET, e.g., E and T, E and ET, or T and ET, where the third and unknown component is calculated as the residual one.

The field measurement of E, T and ET is difficult, particularly early in the growing season when plants are in the seedling growth stage [16] . This has led to the development of simulation models to calculate the partitioning of ET into its two components E and T [8] [16] [41] [49] [51] . The daily calculated ratio of E/ET for the two planting dates is given in Figure 10. As expected in the seedling stage the majority of the evaporative losses are due to E. For example, immediately after irrigation the E/ET ratio was ~70% on DOY 154% and 86% on DOY 153 during the first planting date. The E/ET was 98% on DOY 195 after the field was irrigated and the crop's LAI was 0.07 $m^2{\cdot}m^{-2}$ (Table 1) during the second planting date. The measured cumulative daily ET was 78 mm and 52 mm, respectively for the first and second planting dates. For the first planting date the cumulative E/ET was 45% and 52% for the second planting date. As expected this ratio will decrease as the plants grow and a larger proportion of the ET is due to T. The seasonal E/ET of an irrigated cotton crop in the Texas High Plains was 30% [41] and for a dryland cotton crop the E/ET was 50% and reduced to 31% when the dryland crop was planted in a wheat residue [49] [50].

Examples of the 1/2-hourly record of the mass of a single ML on bare soil, a measure of E, and of a ML with a cotton seedling, a measure of ET, are shown, respectively in Figure 7 and Figure 8. The measurement of 1/2 hourly values of E was for a 12-day period from DOY 161 to 173 (Figure 7). Starting on DOY 161 the initial mass was 12.5 kg at 0850 hours and remained relatively constant until 1800 hours, increasing to 12.9 kg. This increase in ML mass was a result of a 25-mm irrigation applied by the overhead sprinkler (Figure 1(a)). Thereafter, the mass of the ML gradually decreases as a result of water evaporation from the bare soil. The increase in mass on DOY 170.73 (1700 hours) from 12.3 kg to 12.6 kg was again due to a 20-mm irrigation. The example for the 1/2 hourly values of ML mass with a cotton seedling, also show the increase in ML mass due irrigation applied by the center pivot at the end of DOY 161 and on DOY 170.73 (Figure 8).

The average daily values of E and ET measured with the weighable ML's on both planting dates are shown in Figure 9. The gaps in the data were due to rainfall that precluded access to the plot to weigh ML's with the portable field scale. The values of E ranged from a low of ~0.5 $mm{\cdot}d^{-1}$ on several days (DOY 158, 166 - 167, and DOY 208 - 209) to a high of 9.2 $mm{\cdot}d^{-1}$ on DOY 156 and 9.5 $mm{\cdot}d^{-1}$ on DOY 163 for the first planting date and a high of 7.2 $mm{\cdot}d^{-1}$ on DOY 195 for the second planting date. These high values of E can be expected on the Amarillo soil series, particularly after an irrigation or rainfall event where a high rate of E is followed by a much lower value immediately afterwards, i.e., 9.5 $mm{\cdot}d^{-1}$ on DOY 163 to 2.3 $mm{\cdot}d^{-1}$ on DOY 164, and 1.5 mm on DOY 165 [48] [49] . The daily values of ET ranged from a low of ~0.5 $mm{\cdot}d^{-1}$, measured on several days on both planting dates, to high values of 12.8 $mm{\cdot}d^{-1}$ on DOY 156 and 11.1 $mm{\cdot}d^{-1}$ on DOY 163. However, the majority of ET was due to a high value of E. The ability to separate E from ET provides an estimate of plant transpiration (T) values, which are low early in the growing season and increase as plants grow and develop [41] [49] .

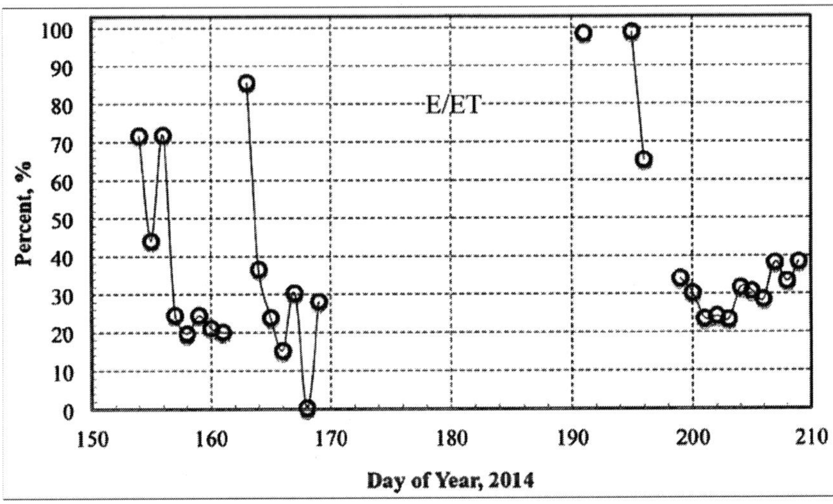

Figure 10. The ratio of daily soil water evaporation (E) to cotton seedling evapotranspiration (ET) in percent for the duration of the two planting studies.

Information on evaporative losses of water from the soil and the plant, throughout the growing season, is an important subject in cropping systems research for both dryland and irrigated crops [41] [50] . In managing water from either rain or irrigation, a frequent goal is to minimize the evaporative losses of water from the soil and to increase crop transpiration. This requires information on two of three components of ET, e.g., E and T, E and ET, or T and ET, where the third and unknown component is calculated as the residual one. The field measurement of E, T and ET is difficult, particularly early in the growing season when plants are in the seedling growth stage [16] . This has led to the development of simulation models to calculate the partitioning of ET into its two components E and T [8] [16] [41] [49] [51] . The daily calculated ratio of E/ET for the two planting dates is given in Figure 10. As expected in the seedling stage the majority of the evaporative losses are due to E. For example, immediately after irrigation the E/ET ratio was ~70% on DOY 154 and 86% on DOY 153 during the first planting date. The E/ET was 98% on DOY 195 after the field was irrigated and the crop's LAI was 0.07 $m^2 \cdot m^{-2}$ (Table 1) during the second planting date. The measured cumulative daily ET was 78 mm and 52 mm, respectively for the first and second planting dates. For the first planting date the cumulative E/ET was 45% and 52% for the second planting date. As expected this ratio will decrease as the plants grow and a larger proportion of the ET is due to T. The seasonal E/ET of an irrigated cotton crop in the Texas High Plains was 30% [41] and for a dryland cotton crop the E/ET was 50% and reduced to 31% when the dryland crop was planted in a wheat residue [49] [50].

4. SUMMARY AND CONCLUSIONS

Our objective was to design and build weighing ML's to measure the ET of cotton seedlings under field conditions. We built ML's made of plastic material that were 35 cm deep, surface area of 181 cm^2 and a volume of 6300 cm^3 that were installed in-situ using a soil-sampling machine that was used to push the plastic tube into the soil to obtain an undisturbed soil core. The soil core was weighed with load-cells that provided a continuous record of the changes in mass due to evaporation of water from the soil and the cotton plant. A change in mass of 18 $g \cdot d^{-1}$ was equivalent to a water evaporation of 1 $mm \cdot d^{-1}$. We compared load-cell measurements of changes in mass to values measured with a portable field scale by linear regression analysis, and the slope was equal to 1, indicating no statistical difference ($P = 0.05$) between the two measurements.

To establish the duration whereby the micro-lysimeters used to measure the seedling ET were representative of cotton plants in the surrounding area, we measured and compared the crop height, leaf area and RL of cotton seedlings in the ML's to cotton plants in the surrounding area. Also, we measured and compared the soil temperature at 10 cm inside ML's with cotton seedlings and the surrounding soil for several days. The measurements of crop height, leaf area and RL suggested that the ML's used were suitable to measure cotton seedling ET for the first 30 days after seed emergence. Also, the measured soil temperature at a 10 cm depth was statistically ($P = 0.05$) the same in and outside the micro-lysimeters. For two planting dates, we measured daily values of E and cotton seedling ET. As expected in the seedling stage, the majority of the evaporative losses are due to E. For example, after an irrigation event, the E/ET ratio was ~70% and 86% the following day. The measured cumulative daily ET was 78 mm for the first planting date and 52 mm for the second planting date. For the first planting date the cumulative E/ET was 45% and 52% for the second planting date. This ratio decreases as the plants grow and a larger proportion of the ET is due to T. Results showed that ML's provide an accurate tool to measure water losses from the soil and cotton plants with a LAI of ≤0.2.

ACKNOWLEDGEMENTS

This research was supported in part by the Ogallala Aquifer Program, a consortium between USDA-Agricultural Research Service, Kansas State University, Texas A & M AgriLife Research, Texas A & M AgriLife Extension Service, Texas Tech University, and West Texas A & M University. We also thank Cotton Incorporated for their support.

REFERENCES

1. Krieg, D.R. (2005) Yield Component Analysis of Cotton—Genetic & Environmental Causes of Variation. Proceedings of 2005 Beltwide Cotton Conferences, New Orleans, 4-7 January 2005, 2022-2032.
2. Krieg, D.R. (2005) Fiber Quality Variation within a Cotton Plant as Affected by Genetics and Environment. Proceedings of 2005 Beltwide Cotton Conferences, New Orleans, 4-7 January 2005, 2380-2385.
3. Guthrie, D., Banks, J.C. and Edmisten, K. (1995) Capitalizing on Earliness. Cotton Physiology Today, Newsletter of the Cotton Physiology Education Program, National Cotton Council, Volume 6, Number 6, 4 p. http://www.cotton.org/tech/physiology/cpt/Growth/upload/Capitalizing-on-Earliness.pdf
4. Robertson, B., Bednarz, C. and Burmester, C. (2007) Growth and Development—First 60 Days. Cotton Physiology Today, Newsletter of the Cotton Physiology Education Program, National Cotton Council, Volume 13, Number 2, 5 p.
5. Gowda, P.H., Baumhardt, R.L., Esparza, A.M., Marek, T.H. and Howell, T.A. (2007) Suitability of Cotton as an Alternative Crop in the Ogallala Aquifer Region. Agronomy Journal, 99, 1397-1403.
6. Oosterhuis, D.M. (1990) Growth and Development of the Cotton Plant. In: Miley, W.N. and Oosterhuis, D.M., Eds., Nitrogen Nutrition in Cotton: Practical Issues, Proceedings of Southern Branch Workshop for Practicing Agronomists, Publications of the American Society of Agronomy, Madison, 1-24.
7. Grimes, D.W. and El-Zik, K.M. (1990) Cotton. In: Stewart, B.A. and Nielsen, D.R., Eds., Irrigation of Agricultural Crops, Agronomy, American Society of Agronomy, Crop Science Society of America, and Soil Science Society of America, Madison, 741-773.
8. Lascano, R.J. (2000) A General System to Measure and Calculate Daily Crop Water Use. Agronomy Journal, 92, 821- 832.
9. Fryrear, D.W. (1971) Survival and Growth of Cotton Plants Damaged by Windblown Sand. Agronomy Journal, 63, 638-642.
10. Krieg, D.R. and Carroll, J.D. (1978) Cotton Seedling Metabolism as Influenced by Germination Temperature, Cultivar, and Seed Physical Properties. Agronomy Journal, 70, 21-25.
11. McMichael, B.L. and Quisenberry, J.E. (1993) The Impact of the Soil Environment on the Growth of Root Systems. Environmental and Experimental Botany, 33, 53-61.

12. Reddy, K.R., Reddy, V.R. and Hodges, H.F. (1992) Temperature Effects on Early Season Cotton Growth and Development. Agronomy Journal, 84, 229-237.
13. Pace, P.F., Cralle, H.T., El-Halawany, S.H.M., Cothren, J.T. and Senseman, S.A. (1999) Drought-Induced Changes in Shoot and Root Growth of Young Cotton Plants. Journal of Cotton Science, 3, 183-187.
14. Baker, J.T. (2007) Cotton Seedling Abrasion and Recovery from Wind Blown Sand. Agronomy Journal, 99, 556-561.
15. Howell, T.A., Schneider, A.D. and Jensen, M.E. (1991) History of Lysimeter Design and Use for Evapotranspiration Measurements. Proceedings of the International Symposium on Lysimetry, Honolulu, 23-25 July 1991, 1-9.
16. Baker, J.T., Gitz III, D.C. and Lascano, R.J. (2014) Field Evaluation of Open System Chambers for Measuring Whole Canopy Gas Exchanges. Agronomy Journal, 106, 537-544.
17. Lascano, R.J. (2007) The Soil-Plant-Atmosphere System and Monitoring Technology. In: Lascano, R.J. and Sojka, R.E., Eds., Irrigation of Agricultural Crops, American Society of Agronomy, Crop Science Society of America, and Soil Science Society of America, Madison, 85-115.
18. Schneider, A.D., Howell, T.A. and Steiner, J.L. (1992) An Evaporation Research Facility Using Monolithic Lysimeters from Three Soils. Applied Engineering in Agriculture, 9, 227-232.
19. Schneider, A.D., Howell, T.A., Moustafa, A.T.A., Evett, S.R. and Abou-Zeid, W. (1998) A Simplified Weighing Lysimeter for Monolithic or Reconstructed Soils. Applied Engineering in Agriculture, 14, 267-273.
20. Marek, T., Piccinni, G., Schneider, A., Howell, T., Jett, M. and Dusek, D. (2006) Weighing Lysimeters for the Determination of Crop Water Requirements and Crop Coefficients. Applied Engineering in Agriculture, 22, 851-856.
21. Evett, S.R., Mazahrih, N.T., Jitan, M.A., Sawalha, M.H., Colaizzi, P.D. and Ayars, J.E. (2009) A Weighable Lysimeter for Crop Water Use Determination in the Jordan Valley, Jordan. Transactions of the ASABE, 52, 155-169.
22. Rogowski, A.S. and Jacoby Jr., E.L. (1977) Assessment of Water Loss Patterns with Microlysimeters. Agronomy Journal, 69, 419-424.

23. Boast, C.W. and Robertson, T.M. (1982) A "Micro-Lysimeter" Method for Determining Evaporation from Bare Soil: Description and Laboratory Evaluation. Soil Science Society of America Journal, 46, 689-696.
24. Boesten, J.J.T.L. and Stroosnijder, L. (1986) Simple Model for Daily Evaporation from Fallow Tilled Soil under Spring Conditions in a Temperate Climate. Netherlands Journal of Agricultural Science, 34, 75-90.
25. Lascano, R.J. and Hatfield, J.L. (1992) Spatial Variability of Evaporation along Two Transects of a Bare Soil. Soil Science Society of America Journal, 56, 341-346.
26. Evett, S.R., Warrick, A.W. and Matthias, A.D. (1995) Wall Material and Capping Effects on Microlysimeter Temperatures and Evaporation. Soil Science Society America Journal, 59, 329-336.
27. Todd, R.W., Evett, S.R., Howell, T.A. and Klocke, N.L. (2000) Soil Temperature and Water Evaporation of Small Steel and Plastic Lysimeters Replaced Daily. Soil Science, 165, 890-895.
28. Martin, E.C., de Oliveira, A.S., Folta, A.D., Pegelow, E.J. and Slack, D.C. (2001) Development and Testing of a Small Weighable Lysimeter System to Assess Water Use by Shallow-Rooted Crops. Transactions of the ASAE, 44, 71-78.
29. Earl, H.J. (2003) A Precise Gravimetric Method for Simulating Drought Stress in Pot Experiments. Crop Science, 43, 1868-1873.
30. Evett, S.R. and Parkin, G.W. (2005) Advances in Soil Water Content Sensing: The Continuing Maturation of Technology and Theory. Vadose Zone Journal, 4, 986-991.
31. Lage, M.A., Bamouh, A., Karrou, M. and El Mourid, M. (2003) Estimation of Rice Evapotranspiration Using a Microlysimeter Technique and Comparison with FAO Penman-Monteith and Pan Evaporation Methods under Moroccan Conditions. Agronomie, 23, 625-631.
32. Ham, J.M., Heilman, J.L. and Lascano, R.J. (1990) Determination of Soil Water Evaporation and Transpiration from Energy Balance and Stem Flow Measurements. Agricultural and Forest Meteorology, 52, 287-301.
33. Baker, J.M. and Spaans, E.J.A. (1994) Measuring Water Exchange between Soil and Atmosphere with TDR-Microlysi- metry. Soil Science, 158, 22-30.

34. Plauborg, F. (1995) Evaporation from Bare Soil in a Temperate Humid Climate-Measurement Using Microlysimeters and Time Domain Reflectometery. Agricultural and Forest Meteorology, 76, 1-17.
35. Zhang, Z., Wang, X., Li, X. and Zhang, J. (2005) Soil Evaporation in Artificially Re-Vegetated Desert Area. Journal of Desert Research, 25, 243-248.
36. Dugas, W.A. and Bland, W.L. (1989) The Accuracy of Evaporation Measurements from Small Lysimeters. Agricultural and Forest Meteorology, 46, 119-129.
37. Heusinkveld, B.G., Berkowicz, S.M., Jacobs, A.F.G., Holtslag, A.A.M. and Hillen, W.C.A.M. (2006) An Automated Microlysimeter to Study Dew Formation and Evaporation in Arid and Semiarid Regions. Journal of Hydrometeorology, 7, 825-832.
38. Baumhardt, R.L., Lascano, R.J. and Evett, S.R. (2000) Soil Material, Temperature, and Salinity Effects on Calibration of Multisensor Capacitance Probes. Soil Science Society of America Journal, 64, 1940-1946.
39. McMichael, B. and Lascano, R.J. (2003) Laboratory Evaluation of a Commercial Dielectric Soil Water Sensor. Vadose Zone Journal, 2, 650-654.
40. Evett, S.R. and Parkin, G.W. (2005) Advances in Soil Water Content Sensing: The Continuing Maturation of Technology and Theory. Vadose Zone Journal, 4, 986-991.
41. Lascano, R.J., Van Bavel, C.H.M., Hatfield, J.L. and Upchurch, D.R. (1987) Energy and Water Balance of a Sparse Crop: Simulated and Measured Soil and Crop Evaporation. Soil Science Society of America Journal, 51, 1113-1121.
42. Evett, S.R., Tolk, J.A. and Howell, T.A. (2006) Soil Profile Water Content Determination: Sensor Accuracy, Axial Response, Calibration, Temperature Dependence, and Precision. Vadose Zone Journal, 5, 894-907.
43. Baker, J.T., Van Pelt, S., Gitz, D.C., Payton, P., Lascano, R.J. and McMichael, B. (2009) Canopy Gas Exchange Measurements of Cotton in an Open System. Agronomy Journal, 101, 52-59.
44. Baker, J.T., Gitz III, D.C. and Lascano, R.J. (2014) Field Evaluation of Open System Chambers for Measuring Whole Canopy Gas Exchanges. Agronomy Journal, 106, 537-544.
45. Newman, E.I. (1966) A Method of Estimating the Total Length of Root in a Sample. Journal of Applied Ecology, 3, 139-145.

46. Holman, D., Sridharan, M., Gowda, P., Porter, D., Marek, T., Howell, T. and Moorhead, J. (2014) Gaussian Process Models for Reference ET Estimation from Alternative Meteorological Data Sources. Journal of Hydrology, 517, 28-35.

47. Baumhardt, R.L., Lascano, R.J. and Krieg, D.R. (1995) Physical and Hydraulic Properties of a Pullman and Amarillo Soil on the Texas South Plains. Technical Report No. 95-1, TAES/TAMU/College Station, College Station, 17 p.

48. Lascano, R.J. and Van Bavel, C.H.M. (1986) Simulation and Measurement of Evaporation from a Bare Soil. Soil Science Society of America Journal, 50, 1127-1132.

49. Lascano, R.J., Baumhardt, R.L., Hicks, S.K. and Heilman, J.L. (1994) Soil and Plant Water Evaporation from Strip- Tilled Cotton: Measurement and Simulation. Agronomy Journal, 86, 987-994.

50. Lascano, R.J. and Baumhardt, R.L. (1996) Effects of Crop Residue on Soil and Plant Water Evaporation in a Dryland Cotton System. Theoretical and Applied Climatology, 54, 69-84.

51. Kool, D., Agam N., Lazarovitch, N., Heitman, J.T., Sauer, T.J. and Ben-Gal, A. (2014) A Review of Approaches for Evapotranspiration Partitioning. Agricultural and Forest Meteorology, 184, 56-70.

Chapter 3

Recent Updates of the Calibration-Free Evapotranspiration Mapping (CREMAP) Method

Jozsef Szilagyi[1, 2]

[1]Department of Hydraulic and Water Resources Engineering, Budapest University of Technology and Economics, Budapest, Hungary

[2]School of Natural Resources, University of Nebraska-Lincoln, Lincoln, Nebraska, USA

1. INTRODUCTION

This study is the revision of an earlier evapotranspiration (ET) estimation technique [1], called Calibration-Free Evapotranspiration Mapping (CREMAP), applied for Nebraska (Figure 1). Major differences between the current and previous versions are

a. the ET rates of the winter months (December, January, February) are included in the current annual maps;

b. the method became largely automated and thus its application made simpler.

In principle, the recent modeling period could have been extended to include 2010 and 2011, however, the multi-institutional research project [2] that provided the monthly incoming global radiation values was terminated in 2010, thus no radiation data are available after June 2010 from that source. Rather than looking for alternative data sources, the original 10-year long modeling period, i.e., 2000-2009, was kept, thus ensuring that the same data types were employed throughout the study.

In the earlier version of the ET maps it was assumed that the ET rates in the winter time are negligible when one is concerned with the mean annual value. In the light of the present version of ET mapping, this assumption was found true only partially: there are regions within Nebraska for which winter ET indeed seems to be negligible (mostly the north-central and north-western parts) in comparison with water-balance data, while in other regions it is not so. These latter regions include parts of Nebraska (mostly the south and south-west portion) with the highest winter-time daily maximum temperatures and/or with most abundant precipitation (eastern, south-eastern portion of the state). As a result the precipitation (P) recycling ratio (i.e., ET / P) rose from a previously estimated mean annual value of 93% to 95%, leaving an estimated 5% of the precipitation to emerge as runoff (Ro) in the streams.

Naturally, as any estimation method, the current approach is not perfect. In the Pine Ridge Escarpment and in the Niobrara River Breaks regions (Figure 1) the ET estimates (similarly to the previous version of the ET map) had to be corrected via comparison with precipitation data because otherwise they would overestimate ET rates by about 10-20%. The reason is in the gross violation of the underlying hypotheses of the current ET estimation method in areas of very rough terrain. After the corrections in these distinct geomorphic regions, it is believed that the resulting ET rates are quite reasonable across the whole state. Overall, the method yields a state-representative ET rate value of 549 mm/yr, within 2% of the simplified water balance (P – Ro) derived rate of 538 mm/yr, employing the USGS [3] values of computed runoff for catchments with level-8 hydrologic unit codes (HUC), and explains 87% of the observed spatial variance of the water balance ET values among the HUC-8 catchments (there are 70 such watersheds within the state) for the same period.

2. DESCRIPTION OF THE CURRENT ET ESTIMATION METHOD

An ET estimation method had been proposed by Bouchet [4], employing the complementary relationship (CR) of evaporation which was subsequently formulated for practical regional-scale ET applications by Brutsaert & Stricker [5] and Morton et al. [6]. In this study the WREVAP program of [6] was applied for the estimation of the regional-scale ET rates at monthly periods. Disaggregation of the regional ET value in space is based on the Moderate Resolution Imaging Spectroradiometer (MODIS) data [7] that have a nominal spatial resolution of about 1 km. The disaggregation is achieved by a linear transformation of the 8-day composited MODIS daytime surface temperature (T_s) values into actual ET rates on a monthly basis [1, 8] by first aggregating the composited T_s data into monthly mean values. Compositing is used for eliminating cloud effects in the resulting composite data by removing suspicious, low pixel-values in the averaging over each eight-day period. See [7] for more detail of data collection and characteristics.

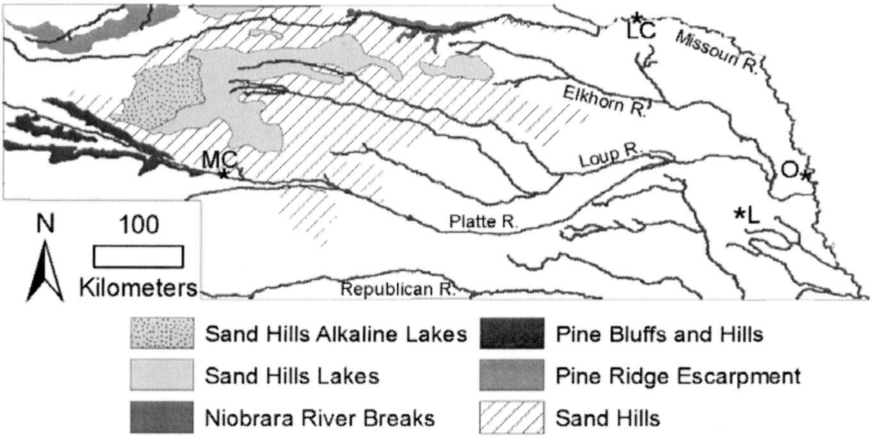

Figure 1. Stream network and selected geomorphic regions of Nebraska. MC: McConaughy Reservoir; LC: Lewis and Clark Reservoir; L: Lincoln; O: Omaha.

The transformation requires the specification of two anchor points in the T_s versus ET plane (Figure 2). The first anchor point is defined by the spatially averaged daytime surface temperature, $\langle T_s \rangle$, and the corresponding regionally representative ET rate, E, from WREVAP. (The original FORTRAN source code can be downloaded from the personal website of the author: snr.unl.edu/ szilagyi/szilagyi.htm). The second anchor point comes from the surface temperature, T_{sw}, of a completely wet cell and the corresponding wet-environment evaporation, E_w, (defined by the Priestley-Taylor [9] equation with a coefficient value of 1.2). The two points identify the linear transformations of the T_s pixel values into ET rates for each month. The resulting line is extended to the right, since in about half the number of the pixels ET is less than the regional mean, E. A monthly time-step is ideal because most of the watershed- or large-scale hydrologic models work at this time-resolution, plus a monthly averaging further reduces any lingering cloud effect in the 8-day composited T_s values. Wet cells within Nebraska were identified over Lake McConaughy and the Lewis and Clark Lake on the Missouri River (Figure 1). An inverse-distance weighting method was subsequently used to calculate the T_{sw} value to be assigned to a given MODIS cell for the linear transformation.

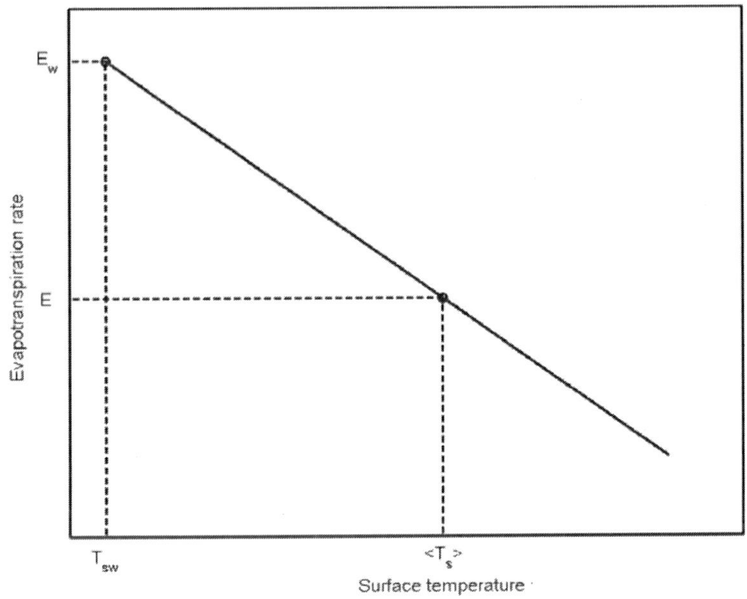

Figure 2. Schematics of the linear transformation of the MODIS daytime surface temperature values into ET rates (after [1]), applied in CREMAP.

The core assumption of CREMAP is that the surface temperature of any MODIS cell is predominantly defined by the rate of evapotranspiration due to the large value of the latent heat of vaporization for water and that the energy (Q_n) available at the surface for sensible (i.e., heat convection) and latent heat (i.e., evapotranspiration) transfers are roughly even among the cells of a flat-to-rolling terrain. Heat conduction into the soil is typically negligible over a 24-hour period, and here considered negligible over the daytime hours as well. This last assumption is most likely true for fully vegetated surfaces where soil heat conduction is small throughout the day, and is less valid for bare soil and open water surfaces. While a spatially constant Q_n term at first seems to be an overly stringent requirement in practical applications due to spatial changes in vegetation cover as well as slope and aspect of the land surface, Q_n will change only negligibly in space provided the surface albedo (i.e., the ratio of in- and outgoing short-wave radiation) value also changes negligibly among the pixels over a flat or rolling terrain. For the study region, the MODIS pixel size of about 1 km may indeed ensure a largely constant Q_n value among the pixels since the observed standard deviation in the mean monthly (warm season) surface albedo value of 17% is only 1.2% among the MODIS cells.

A further assumption of the method is that the vertical gradient of the air temperature near the surface is linearly related to the surface temperature [10,

11], thus sensible heat (H) transfer across the land-atmosphere interface, provided changes in the aerodynamic resistance (r_a) among the MODIS pixels are moderate, can also be taken a linear function of T_s. This can be so because under neutral atmospheric conditions (attained for time-steps a day or longer) r_a depends linearly on the logarithm of the momentum roughness length, z_{0m} [11], thus any change in z_{0m} between pixels becomes significantly dampened in the r_a value due to the logarithm. Consequently, the latent heat (LE) transfer itself becomes a linear function of T_s under a spatially constant net energy (Q_n) term required by the CR, therefore $Q_n = H + LE$, from which $LE = mT_s + c$ follows, m and c being constants for the computational time step, i.e., a month here, within a region.

8-day composited MODIS daytime surface temperature data were collected over the 2000 – 2009 period. The 8-day composited pixel values were averaged for each month to obtain one surface temperature per pixel per month, except for December, January, and February. The winter months were left out of the linear transformations because then the ground may have patchy snow cover which violates the constant Q_n assumption since the albedo of snow is markedly different from that of the land. Therefore in the wintertime the WREVAP-derived regional ET rates were employed without any further disaggregation by surface temperatures but, rather, with a subsequent correction, discussed later.

Mean annual precipitation, mean monthly maximum/minimum air and dew-point temperature values came from the PRISM database [12] at 2.5-min spatial resolution. Mean monthly incident global radiation data at half-degree resolution were downloaded from the GCIP/SRB site [2]. While previously [1] the regions were defined by subdividing the state into eight distinct areas (a largely arbitrary process) for the calculation of the regionally representative values of the mean monthly air temperature, humidity and radiation data, required by WREVAP, now such a subdivision is not necessary. Instead, a "radius of influence" is defined over which the regional values are calculated separately for each designated MODIS cell, very similar to a temporal moving-average process, but now in two dimensions of space. In principle, such a spatial averaging could be performed for each MODIS cell, in practice however, it becomes computationally overwhelming on the PC, and it is also unnecessary, since the spatial averages form a 2-D signal of small gradient, making possible that "sampling" (i.e., the actual calculation of the spatial mean values including the WREVAP-calculated ET rate) is performed only in a selected set of points, which was chosen as each tenth MODIS cell in space (both row-, and column-wise). The remaining cells were then filled up with spatial mean values, linearly interpolated first by row among the selected MODIS-cell values, and then by column, involving the already interpolated values in the rows as well. Near the eastern and southern boundaries of the state any necessary spatial extrapolation was done by the gradient method (i.e., keeping the first two terms of the Taylor-expansion). This "sampling" sped up calculations by at least two orders of magnitude.

Care had to be exercised with the choice of the radius of influence. Rather than applying a constant radius, a spatially changing one was required because near the boundary of the state the "window" becomes asymmetrical around the MODIS cells, therefore the radius changed linearly with distance to these boundaries from a starting value of 25 cells up to a maximum of 125 cells (at a rate of 4/5 cell by each line or column) in the central portion of the domain. It was simpler to define a rectangular region around each designated MODIS cell, rather than a circular one, therefore the radius of influence is half the side-length of the resulting square.

Once the spatial mean values were available for each MODIS cell, the actual linear transformation of the T_s to ET values was performed for each month (except the winter months). The linear transformation of the T_s values into ET rates assumes a negligible change in the r_a value among the cells. As was mentioned above, r_a is directly proportional –up to a constant and with a negative slope— to the logarithm of z_{0m} under neutral stability conditions of the atmosphere, provided the wind speed at the blending height (about 200 m above the ground) is near constant within the region [11]. The momentum roughness height, z_{0m}, of each MODIS cell has been estimated over the state (Figure 3) with the help of a 1-km digital elevation model, as the natural logarithm of the standard deviation in the elevation values among the 25 neighboring cells surrounding a given cell, including the chosen cell itself. The minimum value of z_{0m} has been set to 0.4 m, so when the estimate became smaller than this lower limit (possible for flat regions), the value was replaced by 0.4 m. Note that the z_{0m} = 0.4 m value is the upper interval value for a "prairie or short crops with scattered bushes and tree clumps" in Table 2.6 of [13]. The rugged hills regions of Nebraska (e.g., the Pine Ridge and the Pine Bluffs, just to name a few) are characterized (Figure 3) by the largest z_{0m} values (larger than 3 m), covering the 3-4 m range for "Fore-Alpine terrain (200-300 m) with scattered tree stands" of [13]. Since the r_a estimates are proportional (up to a constant) to the logarithm of the z_{0m} values, their change among the MODIS cells is much subdued: about 67% of the time they are within 5% of their spatial mean and more than 94% of the time they remain within 15% (Figure 4), supporting the original assumption of the present ET mapping method.

Cells that had r_a values smaller than 95% of the mean r_a value (involved about 20% of all cells) were identified, and the corresponding ET values corrected by the relative change in r_a, considering that the sum of the latent (LE) and sensible heat (H) values are assumed to be constant among the cells (equaling Q_n) and that H is proportional to dT_z / r_a [11], where dT_z is the vertical gradient of the air temperature above the surface, itself taken proportional to T_s. The reason that only the "overestimates" of ET are corrected is that the linear transformation of the T_s values into ET rates seems to be more sensitive to more rugged-than-average terrain than to smoother one. That is why the most rugged part of Nebraska, i.e., the Pine Ridge, required an additional (to the above) 10% ET adjustment if no Ponderosa Pine was detected in the 3x3-cell region of the land use-land cover map around a given cell, and a 20% cut if it was. The assumption is that in this extremely rugged region cells with other than Ponderosa pine designation, may still contain scattered trees, if in the vicinity there are pine-forested areas plus the air turbulence, enhanced by the rugged terrain, may have a wake with a characteristic length of about a km. Within the Niobrara River Breaks region (less rugged than the Pine Ridge) only a 10% additional ET adjustment was applied without regard if the cell-surroundings are pine-covered or not. The underlying reasons of these deviations may be (after accepting that the PRISM precipitation values are correct) the way z_{0m} is estimated, perhaps a DEM with a finer resolution would yield better results.

Or maybe the type of vegetation, even at a 1-km resolution, has relevance (similar to plot-scale applications), in addition to the primary elevation variance. Or even, due to the enlarged surface area of the rugged terrain, the global radiation value should be reduced, which would lower ET. This topic certainly requires further research.

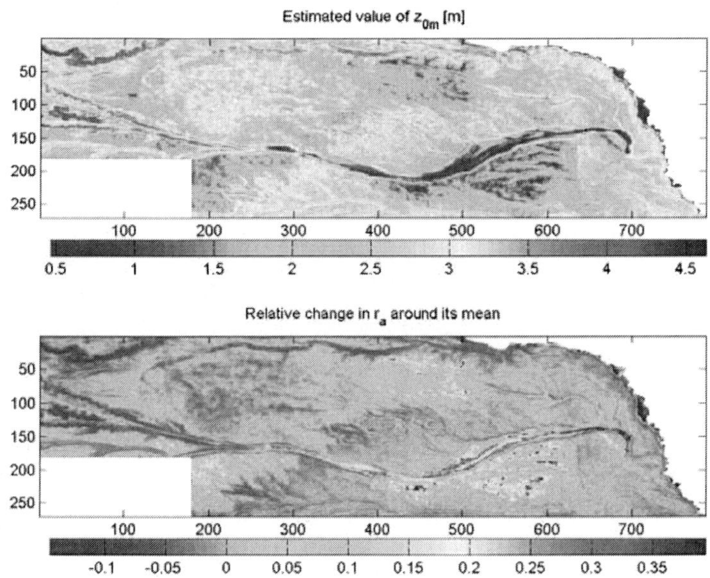

Figure 3. Estimated values of the a) momentum roughness height (z_{0m}), and; b) relative change in the aerodynamic resistance (r_a) around the state-wide mean.

The numbers along the left and bottom edge of the panels are the MODIS cell numbers.

A final correction was applied for cells of "extreme" elevation. Namely, when the elevation of a cell differed from the regional mean value by more than 100 m, its surface temperature was corrected by 0.01 Kelvin per meter, reflecting the dry-adiabatic cooling rate of the air.

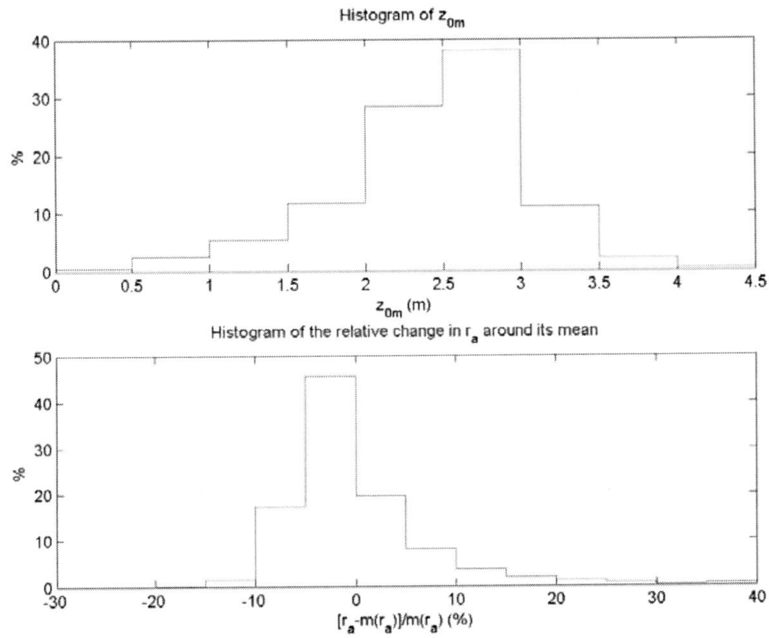

Figure 4. Relative histograms of the momentum roughness height (z_{0m}) and the relative change in the aerodynamic resistance (r_a) around its spatial mean value [$m(r_a)$] across Nebraska, estimated from a 1-km resolution digital elevation model.

The WREVAP model is based on the complementary relationship [4] which performs the worst in the cold winter months [8, 14, 15], thus the resulting WREVAP-obtained winter ET rates become the most uncertain. A yet unpublished study by the present author, involving the Republican River basin, indicated that inclusion of the winter ET rates of WREVAP improved the mean annual ET estimates in comparison with water balance derived [3, 12] data. Other studies [1, 16] also indicated that WREVAP somewhat overestimates ET rates in the Nebraska Sand Hills region even without inclusion of the winter months. Finally, a water balance based [3, 12] verification of the current ET estimates indicated that the WREVAP-provided winter ET rates are necessary in the most humid eastern, south-eastern part of the state. Based on these comparisons, the WREVAP winter months were fully included in the mean annual ET rates [besides the wettest part of the state, defined by (P_{sm} – ET_{WREVAP}) > 50 mm] for the Republican River basin, and for areas where the mean monthly daytime maximum temperature values exceeded 5 ºC. The latter area almost fully covers the Republican River basin, plus the south and south-western part of the panhandle region. P_{sm} designates the spatially smoothed precipitation values of PRISM, applying a 30-by-30-cell window, to filter out the unrealistic grainy structure of the PRISM precipitation field (Figure 5) due probably to its spatial interpolation method. No winter ET rates were included in the mean annual ET values wherever (P_{sm} – ET_{WREVAP}) < 10 mm; and a 50% reduction of the WREVAP winter ET rates were used for areas where [10 mm < (P_{sm} – ET_{WREVAP}) < 50 mm] held true.

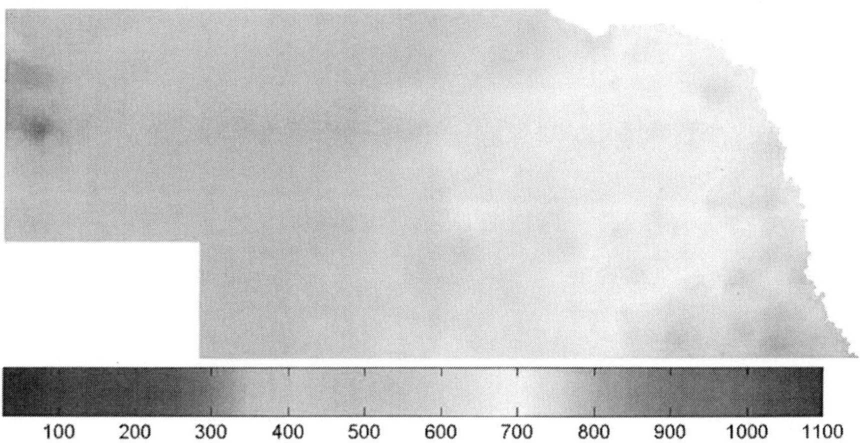

Figure 5. Distribution of the mean annual precipitation (mm) values across Nebraska from the PRISM data, 2000-2009. The state-wide mean annual precipitation rate is 577 mm.

3. RESULTS AND CONCLUSION

The mean annual ET rates across Nebraska are displayed in Figure 6. By and large, ET follows the distribution of precipitation, as expected. Most of the ET values are between 250-500 mm in the panhandle, around 500-650 mm in the middle of the state and near 650 mm in the eastern portion of it. Locally, however, there are large differences due to land use and land cover variance. The sizeable reservoirs (McConaughy, Lewis and Clarke, Harlan County, Swanson, Calamus, etc.) large enough to fully accommodate a MODIS cell, display the largest ET rates, around 1000 mm annually. The reservoirs/lakes are followed by the wider rivers (i.e., Platte, Missouri, Loups, Elkhorn, and partly the Republican) and their valleys in ET rates. The reason, beside the presence of the open water surface, is in the relatively small distance to the groundwater table in these river valleys, enabling the root system of the vegetation to tap into it, plus in the accompanying large-scale irrigation within the valleys. The river valleys on the ET scale are followed by areas of intensive irrigation, reaching 750 mm a year. The driest regions, with the smallest rate of ET in eastern Nebraska are the urban areas of Omaha and Lincoln (Figure 1), where the built in surfaces enhance surface runoff. The eastern outline of the Sand Hills is clearly visible, as well as the sandy areas (the elongated green-colored features) between the Loup and the Platte Rivers. The sandy soil, due to its large porosity favors deep percolation of the water often out of reach of the vegetation.

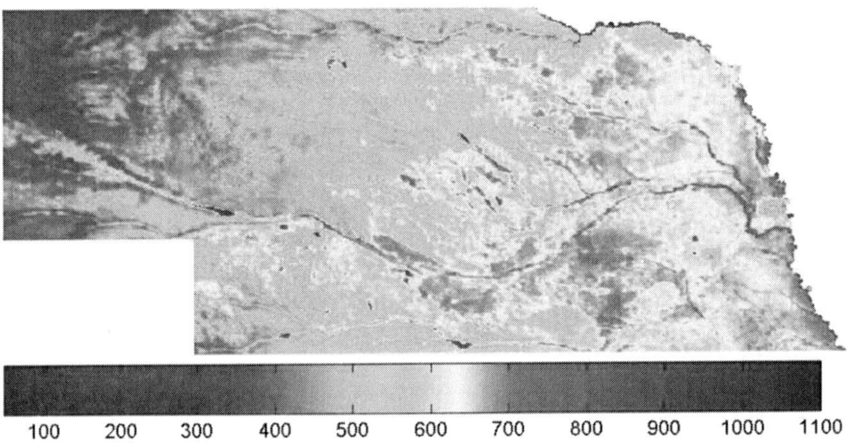

Figure 6. Estimated mean annual ET rates (mm) in Nebraska (2000-2009). The state-wide mean ET value is 549 mm/yr.

Figure 7 depicts the monthly ET rates from January through December. In July and August the irrigated plots in south-central Nebraska can evaporate as intensively as the open water surfaces. Even in September, when most of the produce has been harvested, the soil through its enhanced moisture due to summer irrigation, evaporates more than the surrounding, non-irrigated land. In November the distribution of ET rates becomes zonal and follows the precipitation distribution.

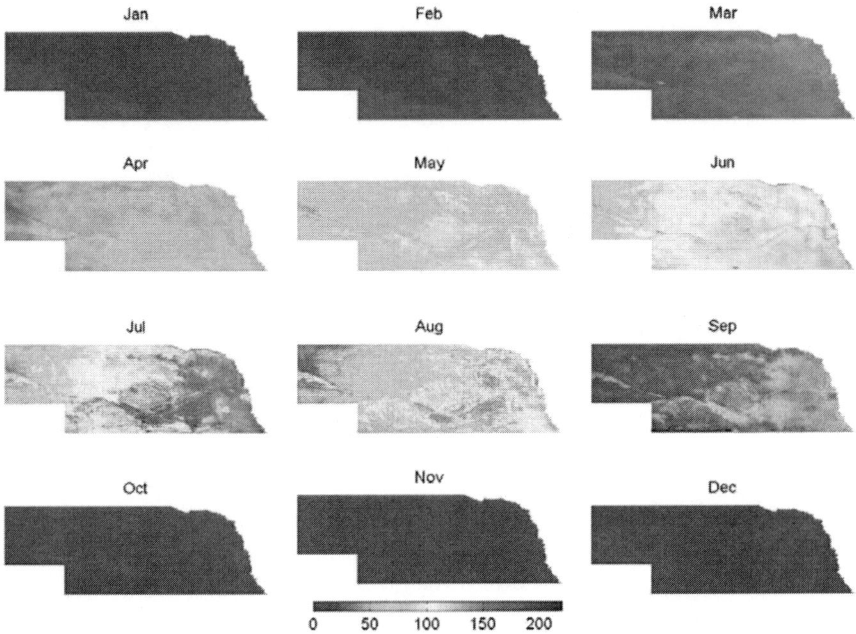

Figure 7. Estimated mean monthly ET rates (mm) in Nebraska., 2000-2009.

While in absolute numbers the south-central portion of the state produces the highest ET rates, the picture changes significantly, when one looks at the ET to precipitation (so-called precipitation recycling) ratios of Figure 8.

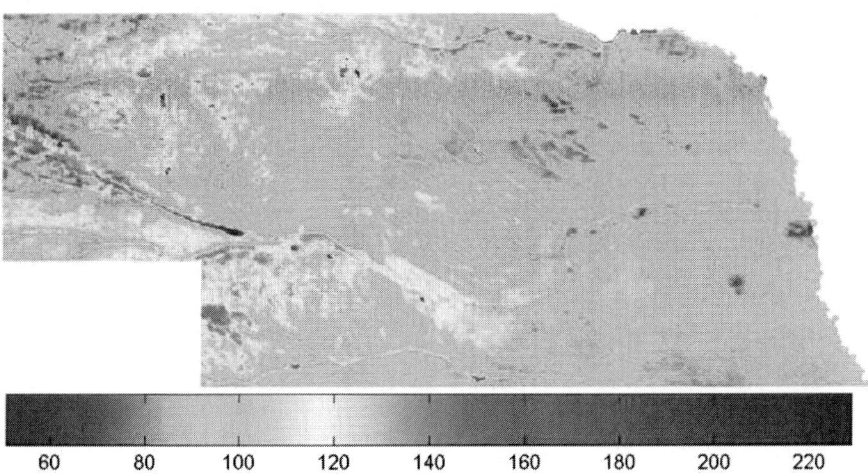

Figure 8. Estimated mean annual ET to precipitation ratios (%), 2000-2009. The state-wide mean ratio is 95%.

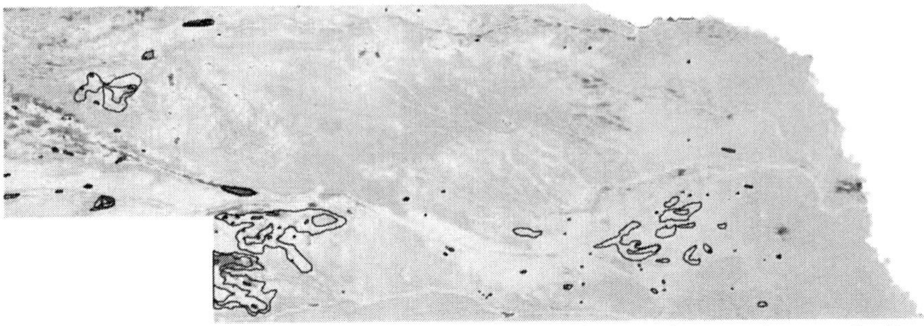

Figure 9. Distribution of areas with the largest observed groundwater decline (at least 3 m, to up to 8 m) over the 2000-2009 period, overlain the ET / P map. For the correct color to ratio correspondence, please, use the colors in Figure 8 instead of the current ones.

Lake McConaughy is the clear winner (followed by smaller lakes in the vicinity), evaporating about twice as much as it receives from precipitation. It does not mean, of course, that the other small lakes in the area would not evaporate as much as Lake McConaughy per unit area, they probably evaporate even more (the smaller the lake the larger typically its evaporation rate, provided other environmental factors are equal), but their size inhibits MODIS to detect their surface temperatures without "contamination" from the surrounding land. Note again the eastern outline of the Sand Hills and the elongated sandy areas between the Loup and Platte Rivers as areas of relatively low ET rates. The two urban areas of Omaha and Lincoln are clearly visible again.

Two large irrigated areas stand out clearly as the most intensive water users (relative to precipitation), one in the Republican River basin and the other in the North-Platte River valley of the panhandle. In these areas ET rates significantly exceed (up to 50%) precipitation rates. Another significant irrigated area in Box Butte County (at the western edge of the Sand Hills) plus the one in the Republican River basin coincide largely with regions of extensive groundwater depletions, displayed in Figure 9.

The lines in Figure 9 designate areas (after [17]) where groundwater decline was at least 3, 5, 8 m over the 2000-2009 period. Naturally, in heavily irrigated areas close to major streams (e.g., North- and South Platte, Platte River), such groundwater depletions are absent (but not around Lake McConaughy, where reservoir water levels have been below normal most of 2000-2009) since the chief source of the irrigation water is the stream itself. Figure 10 displays the distribution of irrigated land, overlain the ET / P values.

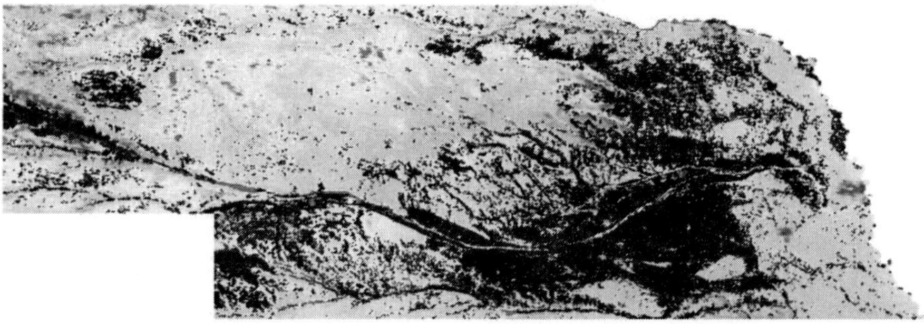

Figure 10. Irrigated land (marked in black) distribution [after 18, 19] in Nebraska, 2005.

As seen in Figure 10, not all land areas with larger than unity ratios are connected to irrigation, good examples are the Sand Hills wetlands. Similarly, not all areas that come up with values larger than 100% do actually evaporate more than they receive from precipitation. Such an area is the table-land just south of the western edge of Lake McConaughy, between the North- and South-Platte Rivers (please, refer to Figure 8 for corresponding precipitation recycling ratios, the colors of Figure 10 are slightly off because it was produced by a different software that enabled marking the irrigated areas on top of the ratios). In this area irrigation is largely absent (or at least it was in 2005, the date of the irrigation data), yet the ratio is between 100-120%. The error may be caused by several factors, namely a) the well-known, often 10% underestimation of precipitation; b) inaccuracy of the ET estimates, and; c) problems with the spatial interpolation of the measured precipitation values. The latter is well demonstrated in Figure 5, which shows that in the southern panhandle region there can be a difference of 125 mm (about 25-30% of the annual value) in the precipitation values within a distance of 30 km or less. Added to this uncertainty is the wide-spread underestimation of precipitation, especially in windy areas where a measurable portion of the raindrops (and especially snowflakes) is swept away from the rain gage. Finally, the present ET estimates have an error term (discussed further later) of about plus/minus (±) 5-10%. Employing a ±5% error in the latter, another -5% underestimation in the measured precipitation values together with yet another independent ±5% error in the interpolated values, the resulting ET / P ratio may contain an error of -5% to 16%, coinciding well with the error extent found in the table-land area. Therefore, the ratios in Figure 8 must be treated with this uncertainty in mind.

A comparison with the previous version [1] of the mean annual ET map (Figure 11) reveals that the largest differences are found in the Republican River basin and the southern panhandle region, where now the full values of the WREVAP-estimated winter-time ET rates were added to the warm-season values (March-November). Note that the procedures used for preparing the two maps are different (application of a radius of influence around each MODIS cell versus distinct geographic regions) as was explained above. The perceptible diagonal and level straight lines suggest some problems with the interpolation method employed in the previous ET map.

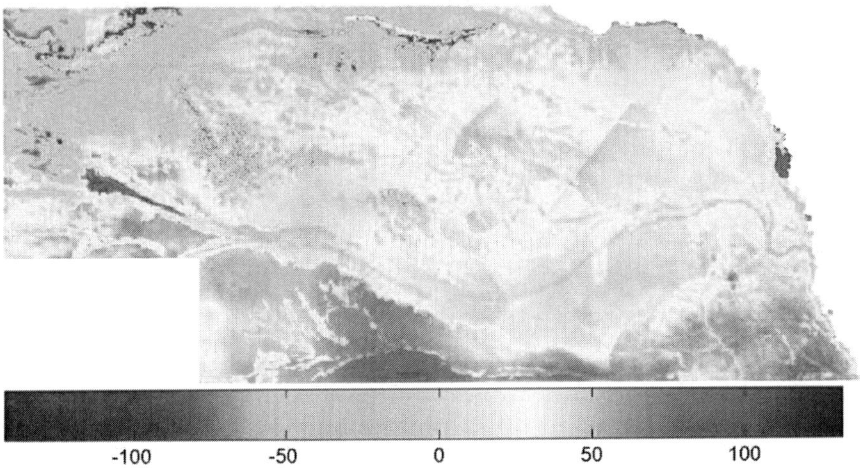

Figure 11. Differences in the present and previous [1] mean annual ET maps (mm). Mean is 18.5 mm.

Verification of the estimated mean annual ET rates can be best performed on a watershed-by-watershed basis by subtracting the stream discharge values (expressed in mm) from the mean precipitation values of the catchments, assuming that groundwater level changes are negligible over the study period, i.e., 2000-2009. As seen above, the latter is not true in many regions within Nebraska, but a transformation of these groundwater-level changes into water depth values would require the state-wide distribution of the specific yield value (also called drainable porosity, defined by dividing the drained water volume value with that of the control volume, fully saturated with water at the start of drainage) of the water bearing aquifer, a hydro-geological parameter not available for the whole state. Figure 12 displays the water-balance derived ET rates employing the PRISM precipitation values [12] and USGS-derived watershed-representative runoff values [3] for the HUC-8 watersheds within Nebraska, while Figure 13 displays the spatial distribution of estimation error (predicted minus water-balance derived) of the CREMAP ET values among the same catchments. As seen, in the majority of the watersheds the estimation error is within 30 mm of the "observed" value. The largest overestimation takes place for watersheds within the Missouri River basin, within the Sand Hills, and within and just north of the Republican River basin. The latter area corresponds to one of the most severe groundwater depletion regions within the state, where the "missing" water certainly contributes to elevated ET rates, not detectable by the simplified water balance approach.

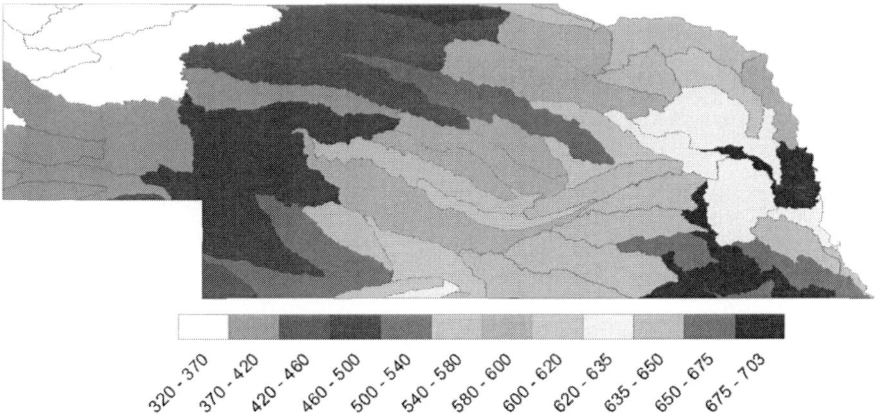

Figure 12. Water balance derived (P – Ro) mean annual ET rates (mm) of the USGS HUC-8 watersheds in Nebraska, 2000-2009.

Note also that a systematic underestimation of the precipitation rates automatically leads to a virtual overestimation of the ET rates by the present method during verification. Another problem with the verification is that the watershed area employed for the transformation of the discharge values into water depth, may also be somewhat uncertain, since the groundwater catchment does not always overlap perfectly with the surface-water catchment delineated by the help of surface elevation values. Probably that is why the largest over- and under-estimation of ET is found within the Sand Hills, in catchments very close to each other. Also, the USGS watershed runoff values employ simplifications that may cause serious errors in the estimated watershed runoff rate, such as found for the Lower Republican basin in Kansas (not shown here), where USGS reports a mean runoff rate of 5 mm/yr for 2000-2009, while a Kansas Geological Survey study [20] found a mean annual runoff rate of 106 mm/yr, an almost twenty-fold difference.

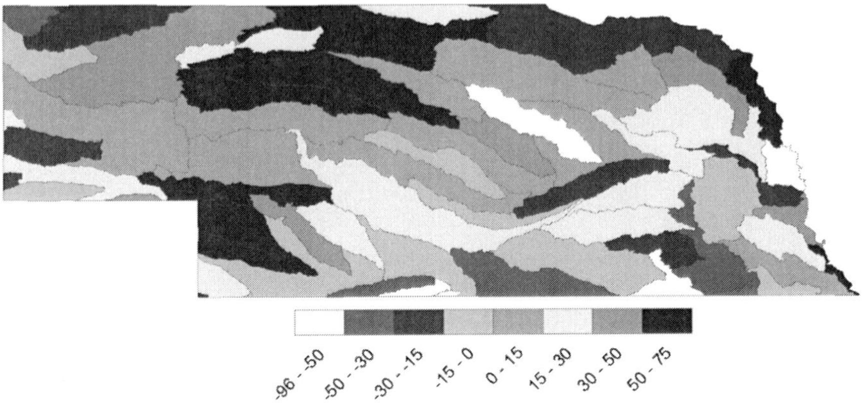

Figure 13. Distribution of the estimation error (mm) in the mean annual ET rates among the USGS HUC-8 watersheds within Nebraska, 2000-2009.

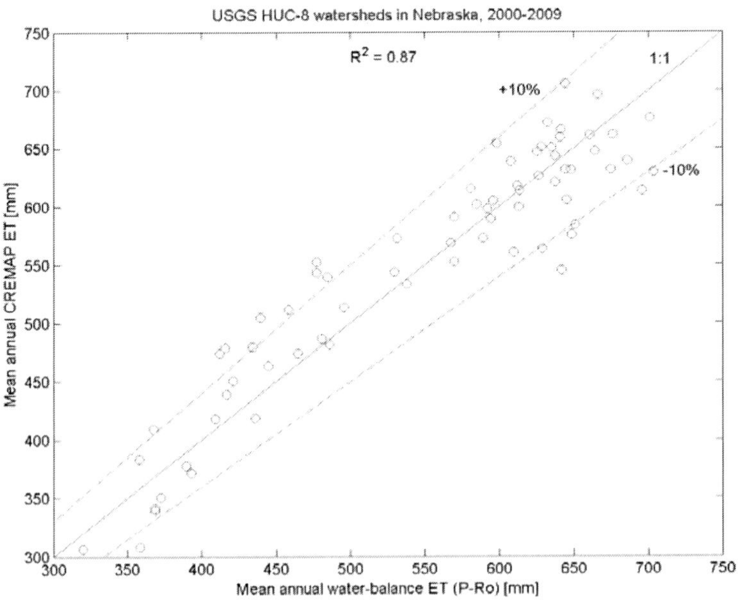

Figure 14. Regression plot of the water-balance derived and CREMAP-estimated mean annual ET rates (mm) among the USGS HUC-8 catchments. R^2 is the portion of the spatial variance of the water balance ET rates that is explained by the CREMAP estimates. The upper and lower envelope lines designate the P – Ro value plus or minus 10%.

Figure 14 summarizes the ET verification results. It can be seen that in the vast majority of the USGS HUC-8 watersheds the estimated values are within 10% of the simplified water balance derived values. Five of the seven overestimates (above the upper intermittent line) of Figure 14, found between 400 and 500 mm, correspond to the large groundwater depletion area in and around the Republican River basin, displayed in Figure 9, so in those cases the CREMAP ET estimates may better represent reality than the simplified water balance derived values. The explained variance, R^2, has a value of 0.87, meaning that 87% of the spatial variance found in the HUC-8 water-balance derived ET values is explained by the CREMAP estimates. In summary, the annual and monthly ET maps are recommended for use in future regional-scale water-balance calculations with the resolution and accuracy of the estimates kept in mind. The maps are certainly not recommended for reading off individual cell values, because the exact cell coordinates maybe slightly off due to the geographically referenced data manipulations necessary to produce those maps. For example, the author found some problem with coordinate referencing when cells are extracted from a grid employing another grid with differing cell size. The maps are best suited for studies of spatial scale larger than one km.

4. ACKNOWLEDGEMENTS

This work has been supported by the Hungarian Scientific Research Fund (OTKA, #83376) and the Agricultural Research Division of the University of Nebraska. This work is connected to the scientific program of the "Development of quality-oriented and harmonized R+D+I strategy and functional model at BME" project. This project is supported by the New Szechenyi Plan (Project ID: TAMOP-4.2.1/B-09/1/KMR-2010-0002).

REFERENCES

1. J. Szilagyi, A. Kovacs, J. Jozsa, 2011 A calibration-free evapotranspiration mapping (CREMAP) technique. In: Labedzki L. (ed.) Evapotranspiration. Rijeka: InTech 257 274Available from http://www.intechopen.com/books/show/title/evapotranspiration (accessed 28 June 2012).
2. GEWEX Continental Scale International Project 2012 Surface Radiation Budget GCIP-SRB. http://metosrv2.umd.edu/~srb/gcip/cgi-bin/historic.cgiaccessed 28 June).
3. United States Geological Survey, USGS. 2012 Computed runoff. http://waterwatch.usgs.gov/new/accessed 28 June).

4. R. J. Bouchet, 1963 Evapotranspiration reelle, evapotranspiration potentielle, et production agricole. Annales Agronomique 14 543 824
5. W. Brutsaert, H. Stricker, 1979 An Advection-Aridity approach to estimate actual regional evapotranspiration Water Resources Research 15 443 449
6. F. Morton, F. Ricard, S. Fogarasi, 1985 Operational estimates of areal evapotranspiration and lake evaporation- Program WREVAP. National Hydrological Research Institute (Ottawa, Canada) Paper 24
7. National Aeronautics and Space Administration. 2012 Moderate Resolution Imaging Spectroradiometer MODIS data. http://modis.gsfc.nasa.govaccessed 28 June).
8. J. Szilagyi, J. Jozsa, 2009 Estimating spatially distributed monthly evapotranspiration rates by linear transformations of MODIS daytime land surface temperature data Hydrology and Earth System Science 13 5 629 637
9. C. Priestley, R. Taylor, 1972 On the assessment of surface heat flux and evaporation using large-scale parameters. Monthly Weather Review 100 81 92
10. W. Bastiaanssen, M. Menenti, R. Feddes, A. Holtslag, 1998 A remote sensing surface energy balance algorithm for land (SEBAL): 1. Formulation. Journal of Hydrology 212 198 212
11. R. Allen, M. Tasumi, R. Trezza, 2007 Satellite-based energy balance for mapping evapotranspiration with internalized calibration (METRIC)-model. Journal of Irrigation and Drainage Engineering 133 4 380 394
12. C. Daly, R. P. Neilson, D. L. Phillips, 1994 A statistical-topographic model for mapping climatological precipitation over mountainous terrain Journal of Applied Meteorology 33 140 158Available from www.prism.oregonstate.eduaccessed 28 June 2012).
13. W. Brutsaert, 2005 Hydrology: An Introduction. Cambridge University Press, Cambridge, United Kingdom
14. J. Huntington, J. Szilagyi, S. Tyler, G. Pohll, 2011 Evaluating the Complementary Relationship for estimating evapotranspiration from arid shrublands Water Resources Research 47 W05533
15. J. Szilagyi, M. Hobbins, J. Jozsa, 2009 A modified Advection-Aridity model of evapotranspiration Journal of Hydrologic Engineering 14 6 569 574
16. J. Szilagyi, V. Zlotnik, J. Gates, J. Jozsa, 2011 Mapping mean annual groundwater recharge in the Nebraska Sand Hills, USA Hydrogeology Journal 19 8 1503 1513

17. J. Korus, M. Burbach, L. Howard, 2012 Groundwater-level changes in Nebraska- Spring 2000 to Spring 2010. Conservation and Survey Division, School of Natural Resources, University of Nebraska-Lincoln. http://snr.unl.edu/csdaccessed 26 June).
18. P. Dappen, I. Ratcliffe, C. Robbins, J. Merchant, 2012 Map of 2005 Center Pivots of Nebraska. University of Nebraska- Lincoln http://www.calmit.unl.edu/2005landusestatewide.shtml (accessed 12 June).
19. P. Dappen, I. Ratcliffe, C. Robbins, J. Merchant, 2012 Map of 2005 Land Use of Nebraska. University of Nebraska- Lincoln http://www.calmit.unl.edu/2005landuse/statewide.shtml (accessed 12 June).
20. M. Sophocleous, 2009 Aquifer storage and recovery and the Lower Republican River valley, Kansas. KGS open file report-18, Lawrence, Kansas.

Chapter 4

Estimation of Evapotranspiration and Crop Coefficients of Tendone Vineyards Using Multi-Sensor Remote Sensing Data in a Mediterranean Environment

Silvia Vanino [1], Giuseppe Pulighe [1,*], Pasquale Nino [1], Carlo De Michele [2], Salvatore Falanga Bolognesi [2] and Guido D'Urso [3]

[1]Consiglio per la Ricerca in Agricoltura e l'Analisi dell'Economia Agraria, via Nomentana 41, 00161 Rome, Italy

[2]ARIESPACE S.r.l. Spin Off Company Università degli Studi di Napoli Federico II, Centro Direzionale IS. A3, 80143 Naples, Italy

[3]Department of Agricultural Sciences, University of Naples Federico II, Via Università 100, I-80055 Portici, Italy

ABSTRACT

The sustainable management of water resources plays a key role in Mediterranean viticulture, characterized by scarcity and competition of available water. This study focuses on estimating the evapotranspiration and crop coefficients of table grapes vineyards trained on overhead "tendone" systems in the Apulia region (Italy). Maximum vineyard transpiration was estimated by adopting the "direct" methodology for ET_p proposed by the Food and Agriculture Organization in Irrigation and Drainage Paper No. 56, with crop parameters estimated from Landsat 8 and RapidEye satellite data in combination with ground-based meteorological data. The modeling results of two growing seasons (2013 and 2014) indicated that canopy growth, seasonal and 10-day sums evapotranspiration values were strictly related to thermal requirements and rainfall events. The estimated values of mean seasonal daily evapotranspiration ranged between 4.2 and 4.1 mm·d^{-1}, while midseason estimated values of crop coefficients ranged from 0.88 to 0.93 in 2013, and 1.02 to 1.04 in 2014, respectively. The experimental evapotranspiration values calculated represent the maximum value in absence of stress, so the resulting crop coefficients should be used with some caution. It is concluded that the retrieval of crop parameters and evapotranspiration derived from remotely-sensed data could be helpful for downscaling to the field the local weather conditions and agronomic practices and thus may be the basis for supporting grape growers and irrigation managers.

Keywords: evapotranspiration; crop coefficient; leaf area index; Landsat 8; RapidEye; remote sensing; vineyards; table grapes

1. INTRODUCTION

In semi-arid zones and in the Mediterranean environment, water scarcity and droughts are an increasingly frequent and widespread phenomenon, and a major limitation for viticulture productivity. In particular, in semi-arid regions of southern Europe, table grapes (Vitis vinifera L.) represent a key economic activity [1], where productivity, defined as the ratio between crop produced and water consumed [2,3], is directly associated with the vineyard water consumption and evapotranspiration [4]. In this context, the accurate monitoring of irrigation water management, such as regulated deficit irrigation and partial root-zone drying irrigation, has emerged as a potential way to increase water-use efficiency while maintaining yields of high quality [5,6,7,8,9], with profitable economic benefits for the grape growers linking water and grape prices and quality of grapes [10]. Timely and accurate estimation of table grapes' water requirements are paramount for adequate irrigation scheduling and irrigation planning, to regulate the appropriate balance between vegetative growth, yield, and berry composition, and to save water. Furthermore, in dry areas the enhanced pressure on water resources with other uses requires an improvement of environmental sustainability of grapevine production according to future climate change scenarios [11].

In the past two decades, a large and growing strand of the literature has investigated various methods to calculate vineyard water requirements. Several studies were based either on measurements from weighing lysimeters [6,12,13], or on surface energy balance such as the Eddy Covariance and Bowen ratio methods [14,15,16,17,18]. Overall, these studies indicate significant variation on water requirements and more generally do not provide spatial trends of evapotranspiration, since the measured values are mostly restricted and influenced by a small footprint area [19].

The FAO-56 model is the most commonly used method for the estimation of crop water requirements [20,21]. Crop potential evapotranspiration (ET_p) is estimated as the product of the reference evapotranspiration (ET_0) and the crop coefficient (K_c). The ET_0 value is estimated by meteorological data and represents the effects of weather conditions on the evapotranspiration process [22]. The K_c coefficient is specific to each crop and to the method used for ET_0 [23], and reflects the canopy development and water management practices over the course of the growing season. Consequently, specific values of K_c are necessary to fit different climate regions and soil type conditions, especially considering future climate change scenarios that predict alterations in the growing season period and phenological stages [24,25].

In recent years, several attempts have been made to implement remote sensing data with the FAO-56 model, in order to provide a reliable alternative to the indirect estimation of ET_p and crop coefficient over large scales [4,23,26,27], allowing accurate monitoring of irrigation water management. Remote sensing approaches in the visible and near infrared ranges can be used to determine the crop water requirements in absence of stress, such as occurs for table grapes [28]. In addition, they do not require the development of complex physical models that work with detailed input datasets. In this case, the interpretation of remotely sensed data is based on the known relationships between spectral reflectance and biophysical crop parameters.

A variety of approaches, both empirical and physically-based methodologies, are used for ET_p and K_c estimations. The first type of methods commonly uses the empirical relationships between Vegetation Indices (VI) and field observations to estimate the energy partitioning and canopy parameters like K_c, while the second type of methods relies on the application of radiative transfer models and strong theoretical and physical relationships that reflect the plant water status, crop characteristics, and soil water management [4,29]. A comprehensive review of remotely-sensed Vegetation Indices used to estimate canopy parameters and their advantages and disadvantages was proposed by Glenn et al. and Allen et al. [30,31]. Although physically-based methods outperform those based on a VI approach, the latter methods are simple to deploy and maintain both in time and space [28]. VI widely used to estimate K_c are the Normalized Difference Vegetation Index (NDVI), the Soil Adjusted Vegetation Index (SAVI), and the Weighted Difference Vegetation Index (WDVI). A considerable amount of research has been conducted in recent years to estimate K_c and the temporal evolution for annual and orchard crops under different climate conditions [2,4,26,27,32,33,34,35].

However, very few studies evaluate the crop coefficients of overhead trellis system vineyards, also known in Italian as "tendone" (meaning "big tent") [36], probably due to the fact that it is typical of only some regions in Southern Italy and Southern America. In table grapes, standard K_c values are obtained from the literature in vines trained on a trellis system developed under different climatic conditions. Nevertheless, crop coefficients primarily depend on the dynamics of canopies, light absorption by the canopy, and canopy roughness, which affects turbulence, crop physiology, leaf age, and surface wetness [37]. Moreover, factors determining the K_c of vineyards include the canopy height and architecture of the trellis system, the fraction of the soil covered by the crop, and the presence of mulch [38]. Thus, it is always recommended to adapt the K_c to the local climate, cultivars, and management practices, especially in orchards and vineyards, where the standard parameters may vary considerably from one area to another [26].

Although extensive research has been carried out on the retrieval of crop biophysical variables, there are few studies that have investigated the estimation of K_c from satellite data for vineyards in the Mediterranean. The main objective of this work is to determine evapotranspiration and K_c linked to the Growing Degree Days (GDD) and the main phenological stages of a "tendone" vineyard in the South Italy. The research was carried out during two growing seasons using satellite reflectance data and the FAO-56 single crop coefficient approach to establish a relationship between VI and K_c. The approach described in this paper also constitutes the basis of the irrigation advisory services using Earth Observation (E.O.) data, such as the "IRRISAT" service operational in Italy [39].

2. MATERIAL AND METHODS

2.1. Study Area and Vineyard Description

The study area is located in the Apulia region (Figure 1) in the southeastern part of the Italian peninsula, where over 70% of the total area is occupied by agriculture and vineyards cover about 26% of the crop area [40]. The regional topography is mainly flat or slightly sloping, with the exception of the Gargano area, situated in the northwest of the region. The climate of the study area is classified as Mediterranean semi-arid, characterized by moderately cold and rainy winters and dry summer seasons. Mean annual precipitation is about 550 mm, concentrated in the fall and winter; long-term mean air temperature is 15.4 °C, while the minimum and maximum yearly temperatures are 3.5 and 29.5 °C, respectively. The climate, rainfall in particular, exhibits a marked inter-annual variability that makes water availability a permanent threat to the economic development and ecosystem conservation of the region [41]. The soil is typical of the Capitanata region, with a well-differentiated profile, characterized by Quaternary marine sediments of illitic/smectitic composition that originate clay soils with vertic features (cambisols, entisols, vertisols) [42]. Due to its geo-climatic conditions, the region suffers from overall water exploitation, which exposes the water supply system to severe water scarcity events.

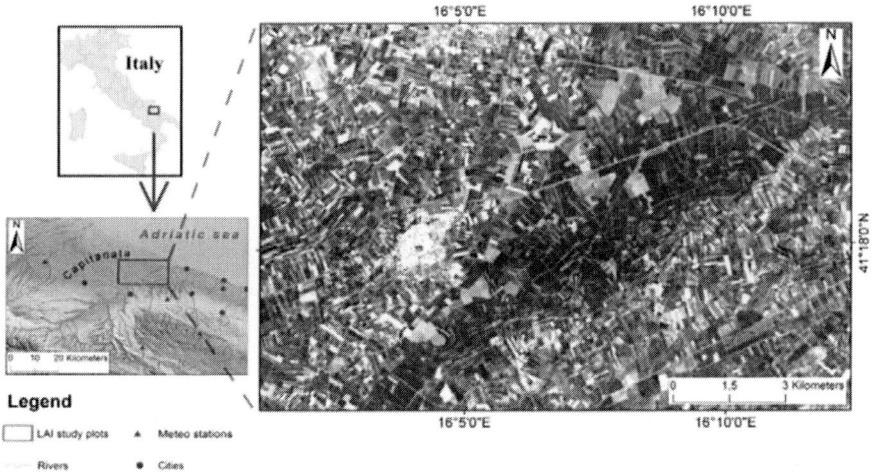

Figure 1. Study area location with overview of LAI plot measurements. Natural color composition of Landsat 8 image acquired on 26 August 2014.

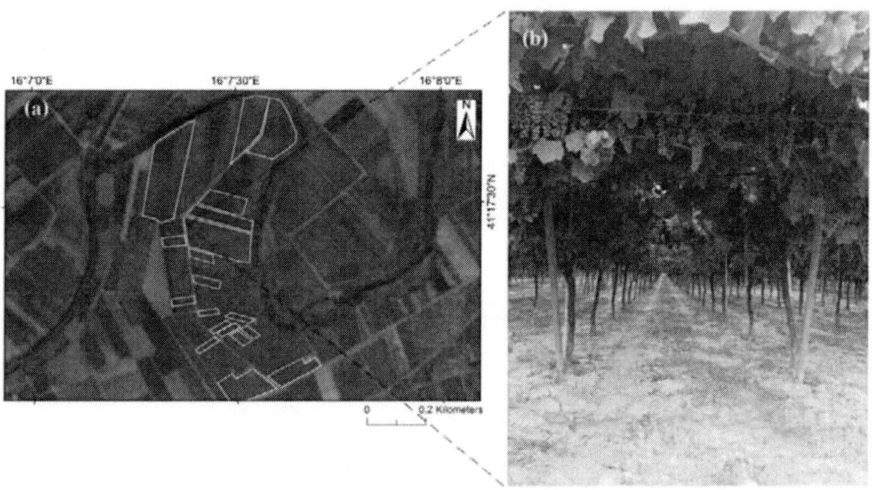

Figure 2. (**a**) Overall view of the vineyard study field displayed on a RapidEye image acquired on 2 August 2014 (infrared false color composition 532). Yellow borders show the experimental plots. (**b**) Canopy growth of the vineyard inter-row acquired on 24 July 2014.

The investigations have been carried within the Land Reclamation and Irrigation Consortium Capitanata, one of the largest agricultural irrigation districts of Southern Italy (Figure 1). This region mainly contains vineyards and orchards over flat terrain. In particular, field data have been collected in a 13 ha multiplot vineyard (41°17'29" North latitude, 16°07'22" West longitude) trained on a "tendone" system, with commercial table grapes (cultivars Italia, Victoria, and RedGlobe), uncovered and drip irrigated (Figure 2). The row and vine spacing in the vineyard was generally 2.3 × 2.3 m, about 2 m high with a plant density of 1890 plants·ha^{-1}. Agronomic management generally maintains the vegetation under non-water stress conditions and soil surface free of weeds and cover crop.

2.2. Dataset

The dataset includes Landsat 8 and RapidEye optical images (in the visible and near infrared ranges), agrometeorological data, and in situ Leaf Area Index (LAI) measurements.

2.2.1. Satellite Images

Earth Observation images used to estimate crop coefficients were collected from the sensor OLI (Operational Land Imager), onboard the Landsat 8 Mission (formerly the Landsat Data Continuity Mission, LDCM). The data are available for download free of charge from the U.S. Geological Survey-USGS Earth Explorer data repository.

In addition, several RapidEye high-resolution images at 6.5 m Ground Sample Distance at nadir (generally resampled to 5-m pixel size), were also acquired. The study field concerns two frames of Landsat 8 images (Path 188, Row 31), radiometrically calibrated and orthorectified using ground control points, and a set of high-resolution RapidEye images, radiometrically calibrated. Table 1 summarizes the cloud-free satellite data used and processed over the study area. The temporal frequency of the entire data set acquired allows us to monitor the vegetation development throughout the growing seasons.

Table 1. Overview of the imagery dataset used in the study area.

Satellite Data	GSD* (m)	Processing Level	Revisit Time	N. of Images	Acquisition Dates
Landsat 8 OLI	30	1T	16 day	14	from 19/05/2013 to 26/08/2014
RapidEye MSI	6.5	1B	1 day	10	from 10/06/2014 to 29/09/2014

Note: *Ground sample distance.

2.2.2. Agrometeorological Data and Field Data

Agrometeorological data were acquired from five stations managed by Assocodipuglia, the agrometeorological service of the Apulia Region. Daily data of precipitation, evapotranspiration, atmospheric pressure, global radiation, air temperature, humidity, and wind speed were acquired for both years. Daily air temperature was used to calculate the GDD index from 1 April to 31 October, computed as the difference between the daily mean temperature and the base temperature (10 °C) [43]. In addition, other ancillary and administrative data were used to improve elaborations.

2.2.3. In Situ LAI

In situ LAI measurements ($m^2 \cdot m^{-2}$) were collected on 23 and 24 July 2014 for 38 sample plots. Field non-destructive measurements of LAI and leaf mean tilt angle were made with LICOR LAI-2000 Plant Canopy Analyzer [44], which works by comparing the intensity of diffuse incident illumination measured at the bottom of the canopy with that arriving at the top. In order to reduce the effect of multiple scattering on LAI-2000 measurements, the instrument was only operated near dusk and dawn (6:30–9:30 am; 6:30–8:30 pm) under diffuse radiation condition, using one sensor for both above and below stand measurements. In order to prevent interference caused by the operator's presence and the illumination condition, the sensor field of view was limited with a 180° view-cap. Measurements were azimuthally oriented opposite to the sun azimuth angle. LAI measurements were taken with the instrument held a few centimeters above the soil, generally within three days of image data acquisition.

A measurement of ambient light was made with the sensor extended upward and over the top of the canopy at arm's length. Eight below-canopy readings were then made. This pattern has been repeated three times per spot, and the resulting 24 samples comprise one full set of measurements. Finally, each center of the LAI-2000 transects was geolocated by using GPS (Global Position System) measurements. This measurement protocol allowed us to determine statistically meaningful LAI values, characterized by a low ratio between the standard error of the LAI (SEL) and LAI (SEL/LAI ranged between ≈0.03 and ≈0.07); thus, the actual LAI should be within 10% of the LAI sample mean (Figure 3).

2.3. Methods

The application of the procedure is achieved by the integration of EO techniques and meteorological data of the study area. The conceptualization has been developed by D'Urso [45] and implemented into the SIMODIS model (SImulation and Management of On-Demand Irrigation System). The estimation of crop evapotranspiration under standard conditions (ET_p—mm/d)—disease-free, adequate fertilization, and soil water availability—has been carried out using the FAO's Penman–Monteith method, which requires standard meteorological data such as solar radiation (S), air temperature (T_a), air humidity (RH), wind speed (U), and biophysical parameters that are crop-specific, such as albedo (r), LAI, and crop height (h_c) [20]. The value of ET_p is a function of the crop parameters and meteorological data, and the procedure, known as the one-step approach, is computed as:

$$ET_p = f\{r, LAI, hc \,|Ta, RH, S, U\} \tag{1}$$

Since crop parameters (mainly r and LAI) are very difficult to obtain in the open field without specific instruments, ET_p is calculated in the routine irrigation management by multiplying ET_0 by K_c, starting from an ET_0 assuming an ideal crop with standard parameters (h_c = 0.12 m, r = 0.23, LAI = 2.88) and a K_c extracted from the FAO-56 table, following a single K_c approach [20].

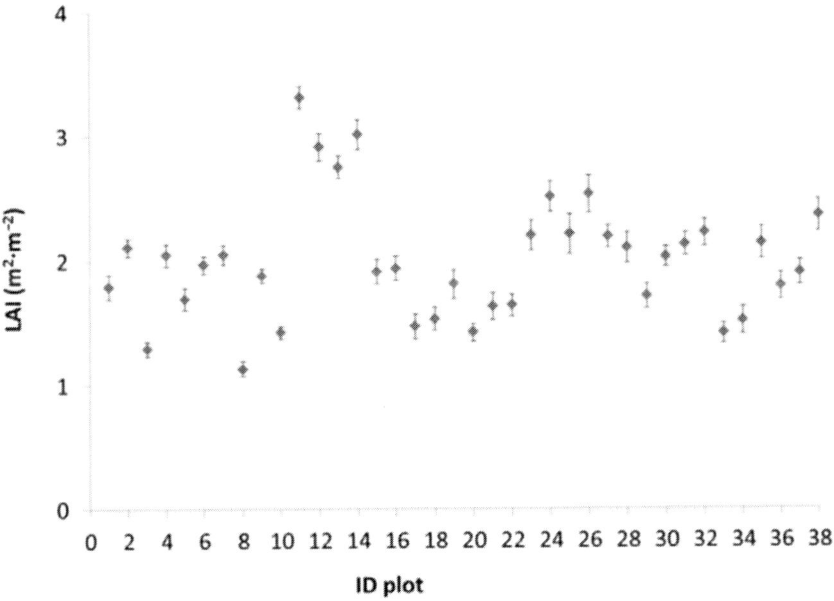

Figure 3. Field estimates of LAI collected on 23–24 July 2014 with a LICOR LAI-2000 Plant Canopy Analyzer. The filled circles show the average LAI with the standard error.

It should be noted that ET_p calculated in this way represents the maximum value, (i.e., minimum canopy resistance), so the resulting K_c values, if compared with values reported in the literature, should be used with caution. The comparison between the results of the proposed methodology and field measurements—including fluxes of latent heat by means of micrometeorological instrumentations—requires fully controlled conditions (especially for irrigation), which were not available at the current study site. However, we assume that the model can be applied to the studied canopy, having been validated in the course of field experiments on different fully irrigated crops [29,39]. Actual evapotranspiration from a well-watered crop will generally approach ET_p during the active growing stage, but it falls below it toward the end of the growing season as the plants begin to dry out. Diversely from our ET_0 estimates, actual evapotranspiration depends on the conditions of soil water availability, and is only possible to compute through complex measurements in the open field or by filling out a detailed hydrological balance, taking into account all the parameters for the simulation of water transport/flow in the Soil, Plant Atmosphere system, using numerical schematization. In this case, the value of ET_p determined with the approach proposed here is used as an upper limiting condition [46]. This kind of analysis has been carried out for some perennial tree crops in Sicily [47], where soil water and surface energy balances with visible, near, and thermal infrared data have been carried out. These previous studies support the assumption of validity, provided that the canopy parameters, i.e., LAI, are adequately described. Finally, crop water requirements can be accounted for in a simple way by subtracting the net precipitation (P_n) from ET_p.

In order to describe the amount of intercepted water from the plant surface, P_n is calculated as a function of the actual precipitation (P), LAI, and fractional cover. The semi-empirical model of interception [48] is described by the following relationship:

$$P_n = P - aLAI\left(1 - \frac{1}{1 + \frac{fcover \cdot P}{aLAI}}\right) \tag{2}$$

where P ($cm \cdot d^{-1}$) and a ($cm \cdot d^{-1}$) are empirical parameters representing the crop saturation per unit foliage area (~0.28 for most crops), and fcover is the fractional vegetation cover derived from LAI using a polynomial empirical expression where coefficients are determined from field measurements and are valid for a wide range of crops ($LAI \leq 5\ m^2 \cdot m^{-2}$).

K_c values are extremely variable, even within the same type of crop, depending on many factors, i.e., date and seeding density, intake of nutrients, nature of the soil, and agronomic practices. K_c is basically the ratio of the ET_p to the ET_0, and can be computed as follows [45,49]:

$$K_c = \frac{ET_p(r^*, LAI^*, hc^*, Ta, RH, U, K^{\downarrow})}{ET_0(r, LAI, hc, Ta, RH, U, K^{\downarrow})} \tag{3}$$

where r^*, LAI^* and h_c^* represent the parameters of the crop vegetation present at the time of the satellite overpass, which can be related to satellite observations, while $K^{\downarrow}$ represent the global incoming short-wave radiation flux density. The value of K_c depends both on the crop parameters (except albedo, which is also influenced by the degree of humidity of the soil), which can be considered relatively stable for a period of 1–2 weeks, and the weather conditions of the study area, which instead vary in time. A sensitivity analysis of the value of K_c with respect to these parameters showed that the value of K_c is more closely related to vegetation parameters r^* and LAI^*, and that the sensitivity of the change in the crop height is higher in autumn than in summer [45]. This is due to the fact that, at the daily scale, the aerodynamic component of evapotranspiration is of much less importance than the radiative. A flow-chart of the processing chain of this procedure can be found in Vuolo et al. [39]. The validation has been carried out over different crops (including vineyards), thereby fulfilling the "standard conditions" for ET_p by means of eddy-covariance independent measurements [29,39]. Since LAI represents the main parameter, it can be taken as the reference variable for the calibration of this method to local conditions, as described later.

Recent studies conducted in semi-arid areas confirm this trend [50]; in fact, a percentage change of 50% in h_c^* corresponds to a variation of 5% of K_c. Therefore, a constant value of $h_c = 0.4$ m has been set in this study, valid for the climatic conditions of the Mediterranean area, reducing the calculation to the estimation of r and LAI.

Data Processing for Deriving EO-Based Crop Development Maps

A total of 14 Landsat 8 images and 10 RapidEye images were used for both growing seasons. The pre-processing phase of satellite images was achieved in two steps. Firstly, Landsat 8 scenes were orthorectified by using additional Ground Control Points and a Digital Elevation Model, while RapidEye scenes were orthorectified using only the image's rational functions. Secondly, the images have been atmospherically corrected using the ATCOR-2 software suite in ERDAS imagine, which incorporates the MODerate resolution atmospheric TRANsmission (MODTRAN version 4) model and uses look-up tables with pre-calculated model simulations for different satellite sensor types and a range of atmospheric conditions [51,52]. Furthermore, supervised classification of Landsat 8 images was performed to identify at each acquisition the plots of agricultural land cover with vegetation and areas with bare soil and/or urban areas.

In order to implement the FAO-56 model, for each satellite image a set of maps of K_c were derived on the basis of LAI and r maps. The surface albedo, needed to derive the net radiant flux, is an approximation of the hemispherical and spectrally integrated surface albedo, and was extracted as part of the value-added products available in ATCOR-2 after the atmospheric correction [53]. LAI is a biophysical surface parameter defined as the total one-sided area of photosynthetic tissue per unit of ground area [54]. The estimate of the LAI was performed based on the simplified Clevers' Leaf Area Index by Reflectance model, CLAIR [55], based on the Weighted Difference Vegetation Index (WDVI). The WDVI is a radiometric index calculated, for each pixel of the image, from the values of reflectance r_{sRED} and r_{sNIR}, respectively, in the bands of red (0.63–0.69 μm) and the near infrared (0.76–0.90 μm). The ratio of near-infrared and red soil reflectance is also known in the literature as the "soil line slope," usually between 0.9 and 1.3. The effect of weighting the red band with the slope of the soil line is the maximization of the vegetation signal in the near-infrared band and the minimization of the effect of soil brightness. The LAI is related with WDVI of the observed vegetation, through the following expression, deduced from a simplified analysis of the radiative behavior of different types of crops [50,55,56]:

$$LAI = -\frac{1}{\alpha^*}\ln(1-\frac{WDVI}{WDVI_\infty}) \tag{4}$$

where α^* is an extinction coefficient (increase of WDVI for an unitary increase of LAI—an empirical shape parameter, mainly depending on canopy architecture and computed from field measurement and considered to be 0.39 in this study, corresponding to the minimum error between the observed and estimated LAI), and $WDVI_\infty$ is the asymptotical value of WDVI for $LAI \rightarrow \infty$, and is calculated at each acquisition at the target "vegetation cover (not woody) very dense and uniform" or maximum vegetation (agricultural) cover (usually between 0.55 and 0.75).

In situ LAI measurements have been used to produce LAI maps from Landsat 8 and RapidEye images by applying the semi-empirical CLAIR model based on an inverse relationship between the WDVI and LAI [55,57].

The CLAIR model needs to be calibrated by estimating the extinction coefficient α^*. Equation (4) must be inverted, fixing the LAI, WDVI, and $WDVI_\infty$ values. $WDVI_\infty$ was calculated (r_{sNIR}/r_{sRED} = 1.35, $WDVI_\infty$ = 0.55) so as to create a correspondence of pixels with maximum vegetation cover. The resulted α^* values were in the range of 0.34–0.70. The final value of α^* = 0.39 corresponds to the minimum error between observed and estimated LAI, leading to an average error of 25% in the estimation of the LAI. To test the model prediction accuracies, we used the root-mean-square error (RMSE) and the coefficient of determination (R^2) between measured and predicted LAI. According to the FAO-56 model, the ET_0 computed from the agrometeorological variables using the Penman–Monteith equation is used to calculate the value of K_c (according to the analytical method), where K_c is the ratio between ET_p and ET_0 [20]. K_c incorporates and synthesizes all the effects on the evapotranspiration related to morpho-physiological characteristics of the different crops—phenological stage, degree of soil cover, soil and climate conditions—that make them different from the reference crop.

The mean value of the ratio between ET_p and ET_0 over a period of about seven days around the date of satellite's passage is applied twice, once with standard values of LAI, r, and crop height for a reference surface, and once by using the LAI, r, and crop height (this latter is fixed for tree crops) from satellite images. To speed up this processing in a GIS environment, we derive a polynomial expression of degree 4, whose coefficients are derived by calculating the K_c intervals of LAI and albedo targets (i.e., LAI in range 0–5 and step 0.2; albedo in range 0.05 to 0.5 and step 0.01). The coefficients mostly depend on meteorological data, and they are dimensionless as LAI and K_c. This procedure is not strictly needed, since we could directly use the FAO-56 method on a daily basis, but it has the advantage of showing the direct link between K_c, LAI, and albedo. Once we derive albedo and LAI from the satellite images and calculate the coefficients for a given time interval, in a GIS environment we derive the K_c map on a pixel-by-pixel basis. One important advantage of deriving canopy parameters or crop coefficients from spectral measurements is that their values do not depend on other variables such as planting date and density, but on the effective cover.

3. RESULTS AND DISCUSSION

3.1. Weather Conditions, Reference ET, and Phenology

Rainfall and temperature in 2013 and 2014 were slightly different but in accordance with climatological averages for the study area (Figure 4). The year 2013 has been characterized by a rainy winter and spring and a dry summer, with anomalously high precipitation in August and in the autumn, while 2014 experienced a rainy spring: long-term average precipitation from April to June was 210 mm, about 35% of the annual amount. This situation affected the crop water requirements during the growing season. Annual rainfall and mean temperature were 629.2 mm and 16.2 °C for 2013, and 607.4 mm and 16.7 °C for 2014, respectively. Total rainfall from 1 June to 30 September was 114.4 mm (18% of the annual amount) in 2013 and 103 mm (27% of the annual amount) in 2014, respectively.

Seasonal variations of ET_0 are shown in Figure 5 for 2013 and 2014, respectively. The total amounts of ET_0 in the growing season (1 June to 30 September) were similar in both years: 511 mm in 2013 and 501.6 mm in 2014. The average daily ET_0 was 4.2 mmd^{-1} and 4.1 mmd^{-1} in the first and second year, respectively. In the 2013 growing season the lowest daily ET_0 was 1.3 mmd^{-1} (30 September) and the highest was 6.7 mmd^{-1} (24 June). In the 2014growing season, ET_0 values ranged from 1.7 mmd^{-1} (1 September) to 6.7 mmd^{-1} (29 June). The transpiration values reported in this study are consistent with those reported by Suvocărev et al. [22] (4.3–4.4 mmd^{-1}) and Teixeira et al. [2] (3.9–4.4 mmd^{-1}) for table grapes trained on an overhead trellis system, although they cannot be directly compared because they were measured at the ground.

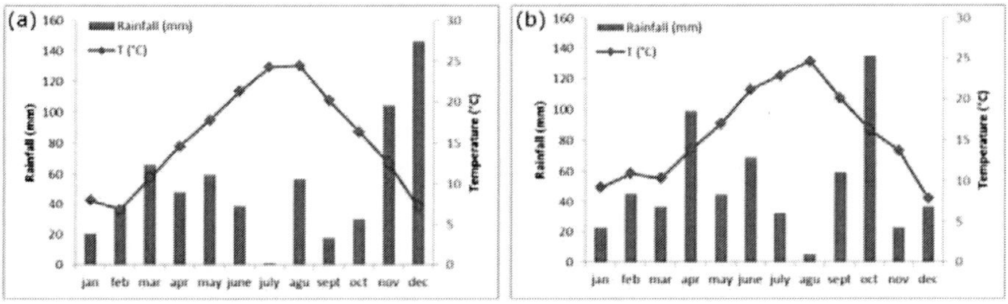

Figure 4. Seasonal trends of temperature and rainfall at an agrometeorological station in Andria for the years 2013 (**a**) and 2014 (**b**).

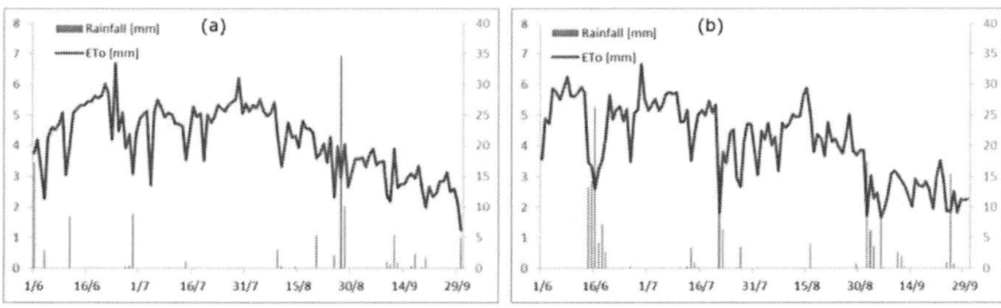

Figure 5. Seasonal trends of ET_0 and rainfall events during the growing season for the years 2013 (**a**) and 2014 (**b**).

In addition, the GDD index was calculated using the data from a weather station located near the sample plots. The length of four key phenological stages was also identified based on mean dates registered in the study area. The GDD index from 1 April to 31 October is similar for both seasons and ranged from 2108 GDD in 2013 to 2128 GDD in 2014 (Table 2), thus the study area was classified as Region IV (hot) [58]. Nevertheless, the GDD accumulation up to flowering in the 2014 growing season was markedly lower than in 2013.

Table 2. Phenological stages and cumulative GDD during the 2013 and 2014 growing seasons.

Phenology *	Time	GDD 2013	GDD 2014
Bud break	1–6 April	0	0
Flowering	20–30 May	341	279
Veraison	16–24 July	981	1040
Harvest	10–15 Sept.	1747	1766
–	31 October	2018	2128

Note: * Observed mean dates of key phenological stages were obtained from the research unit Consiglio per la Ricerca in Agricoltura e l'Analisi dell'Economia Agraria UTV, Turi (Tarricone L., personal communication).

The accumulation of degree-days has important implications for the length of the growth stages and water availability [59]. In fact, it has been demonstrated that canopy development is highly correlated with GDD [12,60] and consequently the canopy cover and shaded area control the seasonal values of the K_c [12]. Overall, these studies suggested that the observed K_c values should be linked to the cumulative GDD of specific agro-climatic locations in order to minimize the effects of year-to-year weather variations. Clearly, the seasonal development and length of growing stages on other study areas can be very different compared with those considered by the FAO-56 paper.

Up to now, a number of studies pointed out that changes in climate led to significant impacts on the phenology time series, for example shifts in phenological timing of bud-burst to harvest [25,61,62]. These results further support the need for setting K_c values for the new climate change projections, as well as agronomic practices aimed at the enhancement of soil water availability (i.e., mulching, netting, and biochar) [11,22,63].

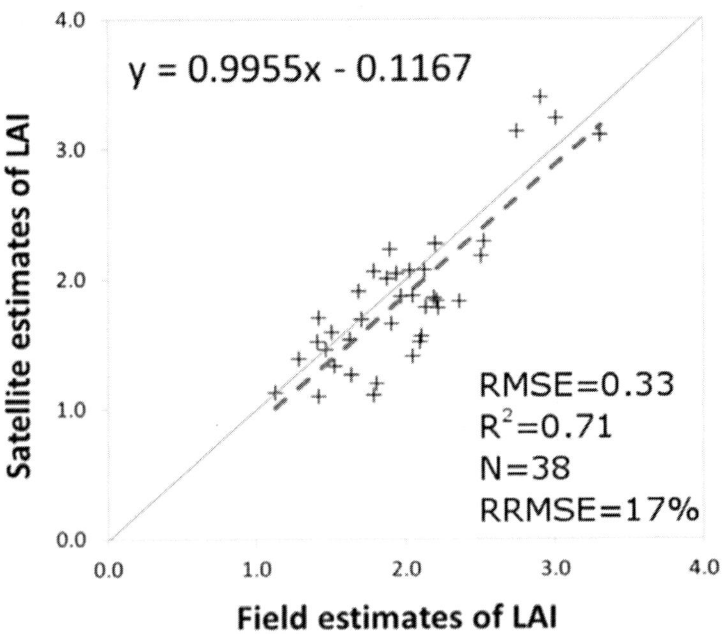

Figure 6. Scatter plots (n = 38) of field LAI estimates (23 to 24 July 2014) and satellite estimates of LAI (RapidEye image, 2 August 2014). RMSE: root mean square error. R^2 = coefficient of determination. RRMSE: relative root mean square error. The broken red line shows the ordinary least squares linear regression fits. In black is the 1-to-1 line.

3.2. Temporal Evolution of LAI and Kc

Consequently, the measurements made in the field were compared with the data elaborated from high-resolution satellite images (RapidEye). The data are an average of the pixels belonging to the same polygon of land cover. The scatterplot between field LAI measured and satellite LAI estimates are shown in Figure 6. The RMSE and R^2 between measured and predicted LAI are 0.33 and 0.71, respectively.

Seasonal K_c values were obtained by dividing ET_p with the reference evapotranspiration calculated as previously described. The time evolution of measured K_c for the growing seasons 2013 and 2014 are reported in Table 3 and Table 4, respectively. During the 2013 growing season the K_c value moves from a minimum of 0.43 to a maximum of 1.23 on the date of harvest, while in 2014 the K_c moves from a minimum of 0.14 to a maximum of 1.42. The K_c mean values were higher in May and June in 2013, and from July to the beginning of September in 2014. Unfortunately, there is no data available for September 2014, since the month was very rainy and the cloud coverage of satellite images did not allow us to elaborate any data. Initial K_c in the 2013 growing season was high and constant around 0.90, with a maximum value of 1.02 close to the harvest, in September. Overall, the standard deviation values of the 2014 data are lower than for the 2013 data, denoting good stability and fairly tight dispersion in the dataset.

Table 3. Statistical parameters for K_c values derived from Landsat 8 satellite images from the year 2013 (n = 110 plots).

K_c	19 May 2013	20 June 2013	06 July 2013	22 July 2013	07 August 2013	23 August 2013	08 Sept 2013	24 Sept 2013
maximum	1.14	1.14	1.14	1.19	1.19	1.18	1.23	1.19
average	0.91	0.91	0.88	0.92	0.93	0.93	1.02	0.96
minimum	0.58	0.58	0.43	0.46	0.48	0.49	0.58	0.56
Standard deviation	0.13	0.13	0.15	0.15	0.14	0.14	0.13	0.13

Table 4. Statistical parameters for K_c values derived from Landsat 8 satellite images from the year 2014 (n = 110 plots).

K_c	06 May 2014	07 June 2014	23 June 2014	09 July 2014	10 August 2014	26 August 2014
maximum	0.63	0.75	1.07	1.42	1.10	1.30
average	0.32	0.43	0.82	1.02	0.93	1.04
minimum	0.14	0.17	0.53	0.68	0.56	0.63
Standard deviation	0.08	0.10	0.11	0.16	0.10	0.12

The review of vineyard K_c values for table grapes available in literature is described in Table 5. Initial K_c in 2014 growing season is 0.32–0.43, closer to the value obtained by Netzer et al. [64] in Israel and Lamaddalena and Caliandro [65] in the same region of our study area. Midseason K_c in our study area in the two growing seasons was around 0.8 and 1.02, in line with those values reported by Rana [36] in Apulia, and Williams and Ayars [12] in California. The peaks of 1.23 in 2013 and 1.42 in 2014 are consistent with those of Netzer et al. [64], Villagra et al. [66], and Williams and Ayars [12]. All these authors estimated a K_c value equal to or greater than 1.2 for table grapevine varieties: "Thompson seedless" for Villagra et al. [66] and Williams and Ayars [12], and "Superior Seedless" for Netzer et al. [64].

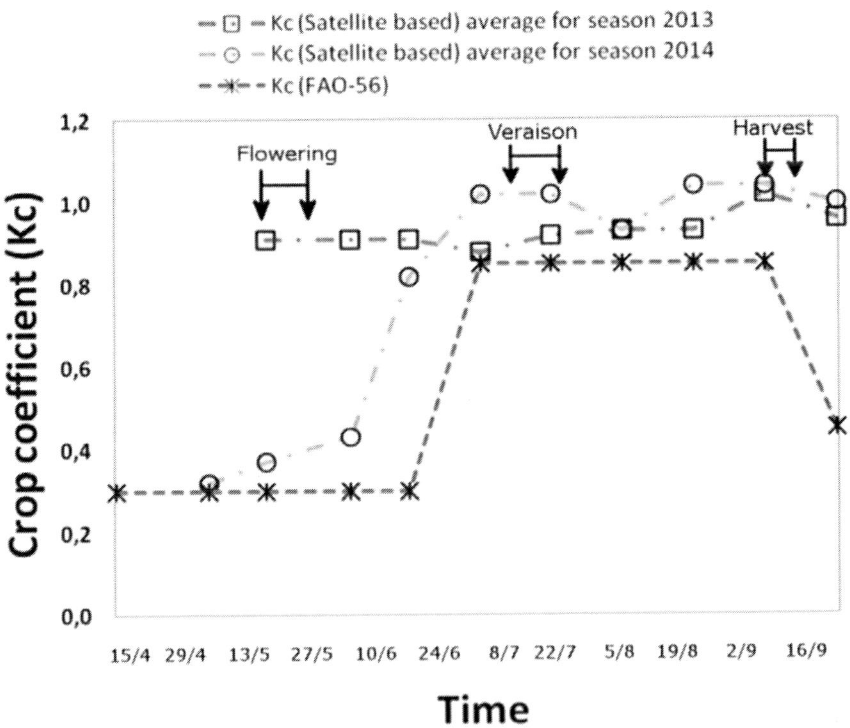

Figure 7. Comparison of Kc (average) estimated from remote sensing data (Landsat 8) and K_c from the FAO-56 table.

Figure 7 shows the temporal evolution of K_c of "tendone" vineyards in 2013 and 2014 together with the K_c of grapes derived from the FAO-56 paper. Furthermore, arrows marked on the graph indicate the corresponding dates of phenological stages. Regarding the 2013growing season, it is interesting to note that the "usual" trend of K_c is not present, probably due to the higher vegetative growth and canopy cover stimulated by high temperatures from bud break up to flowering (Table 2) and regular precipitation. As mentioned in the FAO-56 paper [16], accurate estimates for initial K_c should consider the frequency with which the soil surface is wetted during the initial period. Where the soil is frequently wet from irrigation or rain, and consequently there is a greater amount of available water, the evaporation from the soil surface can be considerable and the initial K_c value will be large.

On the contrary, in the 2014 growing season the trend of estimated K_c value is similar to the K_c FAO trend, with a lower value at the beginning of the season, probably due to low temperatures up to flowering (Table 2), a peak from the first 10 days of July, a slight decrease in the first 10 days of August, and a plateau in the middle of the season. This observation may support the hypothesis that the vegetation growth and K_c trend, after a slow development, was stimulated by GDD accumulation that at veraison surmount the values of the 2013 growing season.

It is interesting to note that the FAO-56 K_c value underestimates the local water need of "tendone" vineyards at all phenological stages. These results are consistent with those of Villagra et al. [66], who reported a K_c value higher than suggested by the FAO-56 paper for Thompson Seedless table grapes in Chile, in growing season 2008–2009 by using the Eddy covariance method. On the contrary, Er-Raki et al. [28] reported K_c values lower than those presented in the FAO-56 paper for Perlette and Superior table grapes in Northwest Mexico (especially for the mid-season) by using ground-based, remotely-sensed data.

However, in the latter case the authors explained their results by the fact that a "Y" open-gable trellis system creates large areas of unshaded bare soil where significant heat fluxes were lost and were not available for transpiration. A possible explanation for these results may be the lack of adequate consideration of vegetation cover, since FAO-56 K_c values were computed for table grapes trained on a trellis system. Furthermore, this critique was confirmed by Williams and Ayars [12], who suggested that K_c and water use are linear functions of the shaded area measured beneath the canopy.

A significant analysis and discussion of this subject was presented by Allen and Pereira [67], who proposed a general procedure for estimating K_c where ground cover and height vary between orchards. In this study the K_c values for the initial, mid-season, and end-season periods were set to 0.30, 1.10, and 0.85 for a ground cover of 70%, 0.30, 0.95, and 0.75 for a ground cover of 50%, and 0.30, 0.60, and 0.5 for a ground cover of 25%.

Overall, these findings suggest that K_c in different crop management systems (e.g., trellising and canopy structure) and agro-climatic locations can vary widely and should be accurately estimated for better irrigation scheduling instead of using the fixed values reported in the literature [12].

Table 5. K_c values of table grapes reported in bibliography.

K_c Initial	K_c Middle	K_c End	Trellis System	Cultivar	Season	Country	References
-	~1	-	Tendone	Italia	1997	Italy	Rana *et al.* [36]
0.2	0.9–1.3	–	Head training system	Thompson seedless	1998–1999	California	Williams and Ayars [12]
0.48	0.68	0.68	Tendone	–	–	Italy	Lamaddalena and Caliandro [65]
0.4	1–1.2	1.3	Open-gable	Superior seedless	2004–2005	Israel	Netzer *et al.* [64]
0.22	0.45	0.3	Open-gable	Perlette & Superior seedless	2005–2006	Mexico	Er-Raki *et al.* [28]
-	0.79	0.98	Open-gable*	Red Globe	2007–2008	Spain	Moratiel and Martinez-Cob [68]
0.54	0.65	0.9	Overhead trellis system*	Crimson seedless	2008–2009	Spain	Suvocarev *et al.* [22]
0.47	0.60	–	Overhead trellis system*	Autumn Royal	2009	Spain	
0.2–0.4	~0.9–1.2	1.2	Overhead trellis system	Thompson seedless	2008–2010	Chile	Villagra *et al.* [66]

Note: * Vineyard covered with a net made of a thread warp of high-density polyethylene.

3.3. Temporal Evolution of Evapotranspiration

Figure 8 provides an overview of the cumulated ET_p and 10-day sums trend line during the period between 1 June and 30 September for both growing seasons. In addition, arrows marked on the graph indicate the dates of the main phenological stages. The average ET_p value ranged from 294 mm in 2013 to 299 mm in 2014. These results can be compared with those obtained by Texeira et al. [2], who reported a mean value ET_p of 372 mm, although, as highlighted by Netzer et al. [64], the comparison between seasonal ET_p in different regions and different practices can be misleading and might lead to misinterpretations of the results and incorrect inferences.

Figure 8 is quite revealing and interesting in several ways, showing differences especially at mid-season. Firstly, the cumulative values for the year 2013 are generally higher than those in year 2014 until the first 10 days of August, while the values have similar trends up to the end of September. As stated previously when referring to the K_c trend lines, these differences can be explained by several factors. First, we need to consider that the 2013 growing season was initially warmer, with enhanced canopy cover and evaporative demand, and consequently improved daily transpiration. Secondly, the 10-day sums of ET_p for both seasons show a trend in line with high atmospheric demand during the summer, but with some peaks and decreasing values linked to rainfall events. For example, for both growing seasons we register the peaks of 10-day sums of ET_p at the days of veraison, but shifted by about 10 days. Furthermore, the peak on the last 20 days of August 2014 might have been influenced by a drought month, while decreasing ET_p values experienced in August 2013 could be attributed to some rain events (about 52 mm in the last 12 days of the month), more abundant and larger than in 2014 (see Figure 4 and Figure 5), that resulted in cloudy and cool conditions. Although mean monthly air temperature was similar for both months (24.4 °C for 2013 and 24.6 °C for 2014), this area is very windy and evapotranspiration trends may also be affected by the wind. Finally, the equal values of ET_p registered toward the end of the season and in correspondence with the harvest can be attributed to reduced evapotranspiration demand and leaf senescence [69]. On the other hand, with regard to the soil water status, it is important to bear in mind that table grape vineyards are constantly irrigated by grape growers to maintain adequate soil water availability to the root-zone.

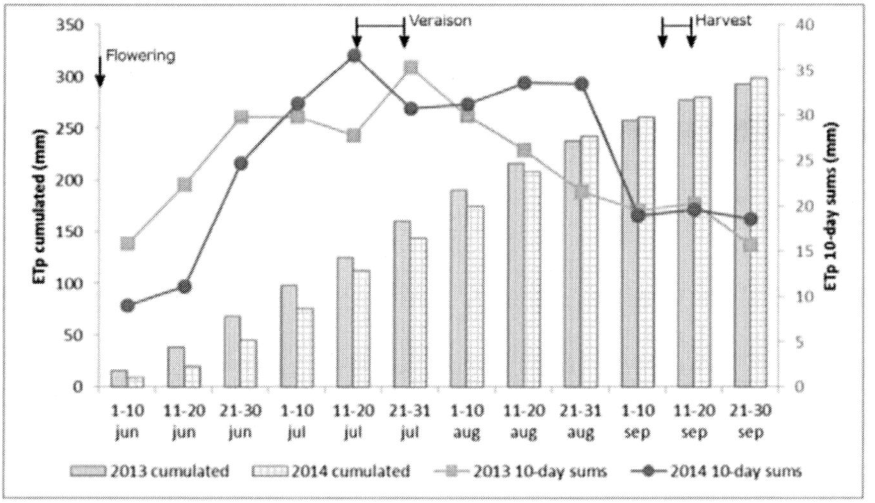

Figure 8. Cumulative values and trend line evolution of ET_p.

Furthermore, crop water requirements calculated during the period June–September as the difference between ET_p and water resulting from precipitation (114.4 mm for 2013 and 164.8 mm for 2014) show an average irrigation requirement of 179.5 mm for 2013 and 134.3 mm for 2014, respectively. The amount of water commonly applied by using drip irrigation in this study area, collected through semi-structured questionnaires engaging grape growers in the 2013season, was approximately 250 mm, close to our ET_p values. Our results indicate that farmers tend to over-irrigate the crop, due to the unavailability of supporting information about climate and canopy development. Interestingly, a recent study in the same study area shows the possibility of reducing crop water requirements by applying seasonal water volumes, equal to 80% of ET_p at the end of irrigation season (about 220–230 mm, in line with our values), with the best balance among grape yield, berry quality, and water use [9]. In addition, visual estimations of time-series in summer 2014 for the LAI, NDVI, K_c, and ET_p are shown in Figure 9. This figure indicates that the ET_p and K_c trend patterns are strictly related with VI at the time of the satellite overpass, but from one date to another are strongly influenced by specific weather conditions and soil moisture. These data show the importance of local climate conditions for water management and irrigation scheduling of the table grapes, and confirm the requirement of site-specific K_c, considering the remarkable seasonality in precipitation during the growing seasons in the study area. Further studies need to be carried out in order to better validate the relation among LAI, K_c, and soil water status.

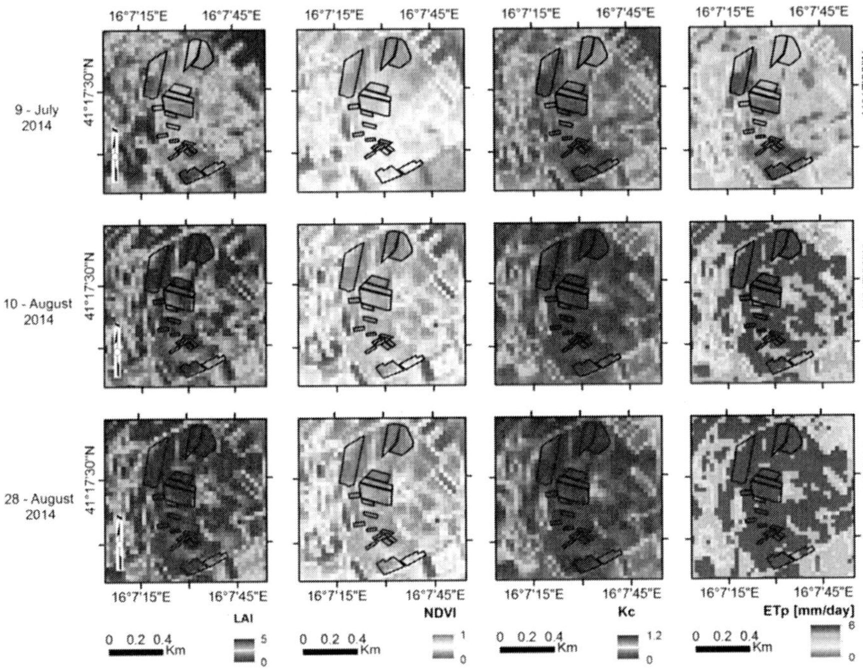

Figure 9. An example of time-series maps of LAI, NDVI, K_c, and ET_p for the vineyard study field represented in Figure 2.

4. CONCLUSIONS

This study evaluates the crop coefficients and evapotranspiration of table grapes trained on a "tendone" system under conditions of unlimited soil water availability, pest and disease-free crop, etc. The method is based on the FAO-56 model by using appropriate values of canopy variables such as the surface albedo and the LAI derived by means of multispectral satellite data, based on existing validated methodologies. The results for the study area in the Apulia region (Italy) indicate that the values of K_c are site- and year-specific and depend on local ET_0 rates, rainfall frequency, cumulative thermal requirements, and effective canopy cover. For instance, during the 2013 growing season the Kc value moves from a minimum of 0.43 to a maximum of 1.23 at harvest, while in 2014 the Kc moves from a minimum of 0.14 to a maximum of 1.42. The obtained results showed good consistency compared with values retrieved from the international literature regarding table grapes in Mediterranean environments, although they should be used with some caution because the calculated ET_p are not the actual values but the maximum values in absence of stress. The resulting crop water requirements are lower than the actual average irrigation volumes applied by farmers, thus suggesting the possibility of improving the water use efficiency of these crops by adopting the proposed methodology. The methodology and results of this research confirm the usefulness of E.O. data in supporting irrigation scheduling and agricultural water management. Furthermore, the approach described in this paper also constitutes the basis for a potential irrigation advisory service using E.O. data as an operational service.

ACKNOWLEDGMENTS

This work has been carried out in the framework of IRMA project (www.irrigation-management.eu), subsidy contract no: I3.11.06. It has been co-financed by EU/ERDF (75%) and national funds of Greece and Italy (25%) in the framework of the European Territorial Cooperation Programme (ETCP) GREECE-ITALY 2007–2013 (www.greece-italy.eu). A special thanks to Enrico Anzano for the valuable work of field data collection with LAI-Licor 2000 and to Luigi Tarricone for the data on phenological stages.

REFERENCES

1. International Organisation of Vine and Wine. Statistical Report on World Viticulture; International Organisation of Vine and Wine: Paris, France, 2013.
2. De C. Teixeira, A.H.; Bastiaanssen, W.G.M.; Bassoi, L.H. Crop water parameters of irrigated wine and table grapes to support water productivity analysis in the São Francisco river basin, Brazil. Agric. Water Manag. **2007**, 94, 31–42.
3. Perry, C.; Steduto, P.; Allen, R.G.; Burt, C.M. Increasing productivity in irrigated agriculture: Agronomic constraints and hydrological realities. Agric. Water Manag. **2009**, 96, 1517–1524.
4. Carrasco-Benavides, M.; Ortega-Farías, S.; Lagos, L.O.; Kleissl, J.; Morales, L.; Poblete-Echeverría, C.; Allen, R.G. Crop coefficients and actual evapotranspiration of a drip-irrigated Merlot vineyard using multispectral satellite images. Irrig. Sci. **2012**, 30, 485–497.

5. Chaves, M.M.; Zarrouk, O.; Francisco, R.; Costa, J.M.; Santos, T.; Regalado, A.P.; Rodrigues, M.L.; Lopes, C.M. Grapevine under deficit irrigation: Hints from physiological and molecular data. Ann. Bot. **2010**, 105, 661–676.
6. Williams, L.E.; Grimes, D.W.; Phene, C.J. The effects of applied water at various fractions of measured evapotranspiration on water relations and vegetative growth of Thompson Seedless grapevines. Irrig. Sci. **2010**, 28, 221–232.
7. Aragüés, R.; Medina, E.T.; Clavería, I.; Martínez-Cob, A.; Faci, J. Regulated deficit irrigation, soil salinization and soil sodification in a table grape vineyard drip-irrigated with moderately saline waters. Agric. Water Manag. **2014**, 134, 84–93.
8. Faci, J.M.; Blanco, O.; Medina, E.T.; Martínez-Cob, A. Effect of post veraison regulated deficit irrigation in production and berry quality of Autumn Royal and Crimson table grape cultivars. Agric. Water Manag. **2014**, 134, 73–83.
9. Tarricone, L.; Gentilesco, G.; Di Gennaro, D.; Amendolagine, A.M. Irrigation strategy and vine performance of organic "Italia" table grape grown in Apulia region (Southern Italy). In Proceedings of the 7th International Table Grape Symposium, Mildura, Victoria, Australia, 11–14 November 2014; pp. 97–100.
10. García García, J.; Martínez-Cutillas, A.; Romero, P. Financial analysis of wine grape production using regulated deficit irrigation and partial-root zone drying strategies. Irrig. Sci. **2012**, 30, 179–188.
11. Medrano, H.; Tomás, M.; Martorell, S.; Escalona, J.-M.; Pou, A.; Fuentes, S.; Flexas, J.; Bota, J. Improving water use efficiency of vineyards in semi-arid regions. A review. Agron. Sustain. Dev. **2014**, 35, 499–517.
12. Williams, L.E.; Ayars, J.E. Grapevine water use and the crop coefficient are linear functions of the shaded area measured beneath the canopy. Agric. For. Meteorol. **2005**, 132, 201–211.
13. López-Urrea, R.; Montoro, A.; Mañas, F.; López-Fuster, P.; Fereres, E. Evapotranspiration and crop coefficients from lysimeter measurements of mature "Tempranillo" wine grapes. Agric. Water Manag. **2012**, 112, 13–20.
14. Li, S.; Tong, L.; Li, F.; Zhang, L.; Zhang, B.; Kang, S. Variability in energy partitioning and resistance parameters for a vineyard in northwest China. Agric. Water Manag. **2009**, 96, 955–962.

15. Li, S.; Kang, S.; Li, F.; Zhang, L.; Zhang, B. Vineyard evaporative fraction based on eddy covariance in an arid desert region of Northwest China. Agric. Water Manag. **2008**, 95, 937–948.
16. Ortega-Farias, S.; Carrasco, M.; Olioso, A.; Acevedo, C.; Poblete, C. Latent heat flux over Cabernet Sauvignon vineyard using the Shuttleworth and Wallace model. Irrig. Sci. **2007**, 25, 161–170.
17. Pellegrino, A.; Gozé, E.; Lebon, E.; Wery, J. A model-based diagnosis tool to evaluate the water stress experienced by grapevine in field sites. Eur. J. Agron. **2006**, 25, 49–59.
18. Zhang, Y.; Kang, S.; Ward, E.J.; Ding, R.; Zhang, X.; Zheng, R. Evapotranspiration components determined by sap flow and microlysimetry techniques of a vineyard in northwest China: Dynamics and influential factors. Agric. Water Manag. **2011**, 98, 1207–1214.
19. Carrasco-Benavides, M.; Ortega-Farías, S.; Lagos, L.; Kleissl, J.; Morales-Salinas, L.; Kilic, A. Parameterization of the Satellite-Based Model (METRIC) for the Estimation of Instantaneous Surface Energy Balance Components over a Drip-Irrigated Vineyard. Remote Sens. **2014**, 6, 11342–11371.
20. Allen, R.G.; Pereira, L.S.; Raes, D.; Smith, M. Crop Evapotranspiration—Guidelines for Computing Crop Water Requirements—FAO Irrigation and Drainage Paper 56; Food and Agriculture Organization of the United Nations: Rome, Italy, 1998.
21. Pereira, L.S.; Allen, R.G.; Smith, M.; Raes, D. Crop evapotranspiration estimation with FAO56: Past and future. Agric. Water Manag. **2014**, 147, 4–20.
22. Suvočarev, K.; Blanco, O.; Faci, J.M.; Medina, E.T.; Martínez-Cob, A. Transpiration of table grape (Vitis vinifera L.) trained on an overhead trellis system under netting. Irrig. Sci. **2013**, 31, 1289–1302.
23. Kamble, B.; Kilic, A.; Hubbard, K. Estimating crop coefficients using remote sensing-based vegetation index. Remote Sens. **2013**, 5, 1588–1602.
24. Moriondo, M.; Jones, G.V.; Bois, B.; Dibari, C.; Ferrise, R.; Trombi, G.; Bindi, M. Projected shifts of wine regions in response to climate change. Clim. Chang. **2013**, 119, 825–839.
25. Moriondo, M.; Ferrise, R.; Trombi, G.; Brilli, L.; Dibari, C.; Bindi, M. Modelling olive trees and grapevines in a changing climate. Environ. Model. Softw. **2015**, 72, 387–401.

26. Campos, I.; Neale, C.M.U.; Calera, A.; Balbontín, C.; González-Piqueras, J. Assessing satellite-based basal crop coefficients for irrigated grapes (Vitis vinifera L.). Agric. Water Manag. **2010**, 98, 45–54.
27. Galleguillos, M.; Jacob, F.; Prévot, L.; French, A.; Lagacherie, P. Comparison of two temperature differencing methods to estimate daily evapotranspiration over a Mediterranean vineyard watershed from ASTER data. Remote Sens. Environ. **2011**, 115, 1326–1340.
28. Er-Raki, S.; Rodriguez, J.C.; Garatuza-Payan, J.; Watts, C.J.; Chehbouni, A. Determination of crop evapotranspiration of table grapes in a semi-arid region of Northwest Mexico using multi-spectral vegetation index. Agric. Water Manag. **2013**, 122, 12–19.
29. D'Urso, G.; Richter, K.; Calera, A.; Osann, M.A.; Escadafal, R.; Garatuza-Pajan, J.; Hanich, L.; Perdigão, A.; Tapia, J.B.; Vuolo, F. Earth Observation products for operational irrigation management in the context of the PLEIADeS project. Agric. Water Manag. **2010**, 98, 271–282.
30. Glenn, E.P.; Huete, A.R.; Nagler, P.L.; Nelson, S.G. Relationship between remotely-sensed vegetation indices, canopy attributes and plant physiological processes: What vegetation indices can and cannot tell us about the landscape. Sensors **2008**, 8, 2136–2160.
31. Allen, R.G.; Pereira, L.S.; Howell, T.A.; Jensen, M.E. Evapotranspiration information reporting: I. Factors governing measurement accuracy. Agric. Water Manag. **2011**, 98, 899–920.
32. Er-Raki, S.; Chehbouni, A.; Guemouria, N.; Duchemin, B.; Ezzahar, J.; Hadria, R. Combining FAO-56 model and ground-based remote sensing to estimate water consumptions of wheat crops in a semi-arid region. Agric. Water Manag. **2007**, 87, 41–54.
33. Er-Raki, S.; Chehbouni, A.; Duchemin, B. Combining satellite remote sensing data with the FAO-56 dual approach for water use mapping in irrigated wheat fields of a semi-arid region. Remote Sens. **2010**, 2, 375–387.
34. Duchemin, B.; Hadria, R.; Erraki, S.; Boulet, G.; Maisongrande, P.; Chehbouni, A.; Escadafal, R.; Ezzahar, J.; Hoedjes, J.C.B.; Kharrou, M.H.; et al. Monitoring wheat phenology and irrigation in Central Morocco: On the use of relationships between evapotranspiration, crops coefficients, leaf area index and remotely-sensed vegetation indices. Agric. Water Manag. **2006**, 79, 1–27.

35. Consoli, S.; Barbagallo, S. Estimating Water Requirements of an Irrigated Mediterranean Vineyard Using a Satellite-Based Approach. J. Irrig. Drain. Eng. **2012**, 138, 896–904.
36. Rana, G.; Katerji, N.; Introna, M.; Hammami, A. Microclimate and plant water relationship of the "overhead" table grape vineyard managed with three different covering techniques. Sci. Hortic. (Amsterdam) **2004**, 102, 105–120.
37. Justice, C.; Townshend, J. Special issue on the moderate resolution imaging spectroradiometer (MODIS): A new generation of land surface monitoring. Remote Sens. Environ. **2002**, 83, 1–2.
38. Fandiño, M.; Cancela, J.J.; Rey, B.J.; Martínez, E.M.; Rosa, R.G.; Pereira, L.S. Using the dual-Kc approach to model evapotranspiration of Albariño vineyards (Vitis vinifera L. cv. Albariño) with consideration of active ground cover. Agric. Water Manag. **2012**, 112, 75–87.
39. Vuolo, F.; D'Urso, G.; De Michele, C.; Bianchi, B.; Cutting, M. Satellite-based Irrigation Advisory Services: A common tool for different experiences from Europe to Australia. Agric. Water Manag. **2015**, 147, 82–95.
40. ISTAT 6 Censimento Generale dell'Agricoltura. Available online: http://dati-censimentoagricoltura.istat.it/ (accessed on 3 July 2015).
41. Vanino, S.; Nino, P.; De Michele, C.; Bolognesi, S.F.; Pulighe, G. Earth Observation for Improving Irrigation Water Management: A Case-study from Apulia Region in Italy. Agric. Agric. Sci. Procedia **2015**, 4, 99–107.
42. Benedetti, A.; Teresa, M.; Abate, D.; Napoli, R. The Soils of Italy; Edoardo, A.C., Costantini, C.D., Eds.; Springer: Dordrecht, Netherlands, 2013.
43. Winkler, A.J.; Cook, J.A.; Kliewer, W.M.; Lider, L.A. General Viticulture, 2nd ed.; University of California Press: Oakland, CA, USA, 1974.
44. LI-COR. LAI-2000 Plant Canopy Analyzer Instruction Manual 1992. Available online: http://www.licor.com/env/products/leaf_area/LAI-2200/ (accessed on 3 July 2015).
45. D'Urso, G. Simulation and Management of on-Demand Irrigation Systems: A Combined Agrohydrological and Remote Sensing Approach. PhD Thesis, Wageningen Agricultural University, Wageningen, Netherlands, 2001.
46. Minacapilli, M.; Iovino, M.; D'Urso, G. A distributed agro-hydrological model for irrigation water demand assessment. Agric. Water Manag. **2008**, 95, 123–132.

47. Minacapilli, M.; Agnese, C.; Blanda, F.; Cammalleri, C.; Ciraolo, G.; D'Urso, G.; Iovino, M.; Pumo, D.; Provenzano, G.; Rallo, G. Estimation of actual evapotranspiration of Mediterranean perennial crops by means of remote-sensing based surface energy balance models. Hydrol. Earth Syst. Sci. **2009**, 13, 1061–1074.
48. Braden, H. Ein Energiehaushalts- und Verdunstungsmodell for Wasser und Stoffhaushaltsuntersuchungen landwirtschaftlich genutzer Einzugsgebiete. M. Dtsch. Bodenkd. Geselschaft **1985**, 42, 294–299.
49. FAO. Irrigation Scheduling: From Theory to Practice. In Water Reports 8, Proceedings of the ICID/FAO Workshop on Irrigation Scheduling, Rome, Italy, 12–13 September 1995; Food and Agriculture Organization of the United Nations: Rome, Italy, 1996.
50. Akdim, N.; Alfieri, S.; Habib, A.; Choukri, A.; Cheruiyot, E.; Labbassi, K.; Menenti, M. Monitoring of Irrigation Schemes by Remote Sensing: Phenology versus Retrieval of Biophysical Variables. Remote Sens. **2014**, 6, 5815–5851.
51. De Vries, C.; Danaher, T.; Denham, R.; Scarth, P.; Phinn, S. An operational radiometric calibration procedure for the Landsat sensors based on pseudo-invariant target sites. Remote Sens. Environ. **2007**, 107, 414–429.
52. Vicente-Serrano, S.; Pérez-Cabello, F.; Lasanta, T. Assessment of radiometric correction techniques in analyzing vegetation variability and change using time series of Landsat images. Remote Sens. Environ. **2008**, 112, 3916–3934.
53. Richter, R. Atmospheric Correction Methodology for Imaging Spectrometer Data. In The Digital Airborne Spectrometer Experiment (DAISEX); Wooding, M., Harris, R.A., Eds.; European Space Agency: Noordwijk, The Netherlands, 2001.
54. Bréda, N.J.J. Ground-based measurements of leaf area index: A review of methods, instruments and current controversies. J. Exp. Bot. **2003**, 54, 2403–2417.
55. Clevers, J.G.P.W. The application of a weighted infrared-red vegetation index for estimating leaf area index by correcting for soil moisture. Remote Sens. Environ. **1989**, 29, 25–37.
56. Vuolo, F.; Neugebauer, N.; Bolognesi, S.F.; Atzberger, C.; D'Urso, G. Estimation of leaf area index using DEIMOS-1 data: Application and transferability of a semi-empirical relationship between two agricultural areas. Remote Sens. **2013**, 5, 1274–1291.

57. Clevers, J.G.P.W. The derivation of a simplified reflectance model for the estimation of leaf area index. Remote Sens. Environ. **1988**, 25, 53–69.
58. Jones, G.V.; Duff, A.A.; Hall, A.; Myers, J.W. Spatial Analysis of Climate in Winegrape Growing Regions in the Western United States. Am. J. Enol. Vitic. **2010**, 61, 313–326.
59. Ramos, M.C.; Martínez-Casasnovas, J.A. Effects of precipitation patterns and temperature trends on soil water available for vineyards in a Mediterranean climate area. Agric. Water Manag. **2010**, 97, 1495–1505.
60. Williams, L.E. Interaction of applied water amounts and leaf removal in the fruiting zone on grapevine water relations and productivity of Merlot. Irrig. Sci. **2012**, 30, 363–375.
61. Poblete-Echeverría, C.A.; Ortega-Farias, S.O. Evaluation of single and dual crop coefficients over a drip-irrigated Merlot vineyard (Vitis vinifera L.) using combined measurements of sap flow sensors and an eddy covariance system. Aust. J. Grape Wine Res. **2013**, 19, 249–260.
62. Tomasi, D.; Jones, G.V.; Giust, M.; Lovat, L.; Gaiotti, F. Grapevine Phenology and Climate Change: Relationships and Trends in the Veneto Region of Italy for 1964–2009. Am. J. Enol. Vitic. **2011**, 62, 329–339.
63. Baronti, S.; Vaccari, F.P.; Miglietta, F.; Calzolari, C.; Lugato, E.; Orlandini, S. Impact of biochar application on plant water relations in Vitis vinifera(L.). Eur. J. Agron. **2014**, 53, 38–44.
64. Netzer, Y.; Yao, C.; Shenker, M.; Bravdo, B.A.; Schwartz, A. Water use and the development of seasonal crop coefficients for Superior Seedless grapevines trained to an open-gable trellis system. Irrig. Sci. **2009**, 27, 109–120.
65. Lamaddalena, N.; Caliandro, A. Riorientamenti Produttivi del Territorio Agricolo Pugliese per uno Sviluppo Rurale Sostenibile: Promozione di Servizi Orientati allo Sviluppo Agricolo—Azione 3; Regione Puglia & CIHEAM Istituto agronomico mediterraneo di Bari: Valenzano, Italy, 2008.
66. Villagra, P.; García de Cortázar, V.; Ferreyra, R.; Aspillaga, C.; Zuñiga, C.; Ortega-Farias, S.; Selles, G. Estimation of water requirements and Kc values of "Thompson Seedless" table grapes grown in the overhead trellis system, using the Eddy covariance metho. Chil. J. Agric. Res. **2014**, 74, 213–218.

67. Allen, R.G.; Pereira, L.S. Estimating crop coefficients from fraction of ground cover and height. Irrig. Sci. **2009**, 28, 17–34.

68. Moratiel, R.; Martínez-Cob, A. Evapotranspiration of grapevine trained to a gable trellis system under netting and black plastic mulching. Irrig. Sci. **2012**, 30, 167–178.

69. De Azevedo, P.V.; Soares, J.M.; Da Silva, V.P.R.; Da Silva, B.B.; Nascimento, T. Evapotranspiration of "Superior" grapevines under intermittent irrigation. Agric. Water Manag. **2008**, 95, 301–308.

Chapter 5

Influence of Potential Evapotranspiration on the Water Balance of Sugarcane Fields in Maui, Hawaii

Javier Osorio[1], Jaehak Jeong[1], Katrin Bieger[1], Jeff Arnold[2]

[1]Blackland Research and Extension Center, Texas A & M AgriLife Research, Temple, USA

[2]Grassland, Soil and Water Research Laboratory, United States Department of Agriculture, Agricultural Research Service (USDA-ARS), Temple, USA

ABSTRACT

The year-long warm temperatures and other climatic characteristics of the Pacific Ocean Islands have made Hawaii an optimum place for growing sugarcane; however, irrigation is essential to satisfy the large water demand of sugarcane. Under the Hawaiian tropical weather, actual evapotranspiration (AET) is the primary mechanism by which water is removed from natural and agricultural systems. The Hawaiian Commercial and Sugar Company (HC&S), the largest sugarcane grower of the Hawaiian Islands, has developed a locally optimized AET equation for the purpose of water management on its 184.3 km^2 sugarcane plantation on the Island of Maui. In this paper, in order to assess the influence of AET on the hydrological water balance of the HC&S' sugarcane cropping system, the performance of the HC&S method was compared with three physically-based methods: Penman-Monteith, Priestley-Taylor, and Hargreaves, as well as, to a set of historical pan evaporation data. A Soil and Water Assessment Tool (SWAT) project was setup to estimate the water balance in two sugarcane fields: a windy lowland field and a rocky highland field on a hill slope. Under Hawaiian weather conditions, wind speed was found to be the most influential climatic parameter over potential evapotranspiration (PET); therefore, the results with both Hargreaves and Priestley-Taylor underpredicted PET by approximately 30%, presumably because these methods do not take wind speed into account. The HC&S method was demonstrated to be the most accurate PET method compared to the other commonly used PET equations, with less than 10% error. Of the annual total water supply of 3400 mm, AET accounted for 75% - 80% of the total water consumption. These findings can be used to improve the irrigation efficiency as well as other management scenarios to optimize water use on the Island of Maui.

Keywords: Evapotranspiration, Water Balance, Hydrological Modeling, Sugarcane, SWAT

1. INTRODUCTION

The year-long warm temperatures and other climatic characteristics of the Pacific Ocean Islands have made the Island of Maui an optimum place for growing sugarcane (Saccharum officinarum L.). Because of the high water demand of sugarcane, irrigation becomes one of Maui's major constraints [1] . According to McMahon et al. [2] irrigation is essential to satisfy the large water demand of sugarcane for optimum growth and yield. In Maui, large quantities of surface and ground water from the windward side of the island are diverted by intake structures into ditches and tunnels to transport the water to the leeward plains, where the best agricultural soils are located [1] . However, mild droughts and periods of low rainfall have adversely affected the perennial streams and depleted high-level groundwater aquifers that supply Maui's irrigation system. As water resources become limited, effective use of available water for irrigation is of concern throughout the sugarcane industry [3] .

The Hawaiian Commercial & Sugar Company (HC&S), the largest sugarcane grower of the Hawaiian Islands, has an area of 184.3 km^2 under sugarcane production on the Island of Maui. The productivity of sugarcane on Maui (159.9 $t \cdot ha^{-1}$ sugarcane yield and 24.9 $t \cdot ha^{-1}$ sugar yield) is known to be above the national average (78.8 $t \cdot ha^{-1}$ sugarcane yield and 9.2 $t \cdot ha^{-1}$ sugar yield) and higher than other major sugarcane producing regions in USA like Louisiana (51.3 $t \cdot ha^{-1}$ sugarcane yield and 6.6 $t \cdot ha^{-1}$ sugar yield) or Texas (85.9 $t \cdot ha^{-1}$ sugarcane yield and 10.0 $t \cdot ha^{-1}$ sugar yield) [4] . The sugarcane cropping period is approximately 24 months. The company requires 760,000 $m^3 \cdot d^{-1}$ of fresh water for irrigation, which is obtained from surface (fog drip and rainfall) and ground water sources (16 deep brackish-water wells) and is applied directly to the root zone of the sugarcane plants through a drip irrigation system.

From the hydrological perspective, actual evapotranspiration (AET) is the primary mechanism by which water is removed from the hydrologic cycle [5] - [7] . AET returns land-based water to the atmosphere through processes such as transpiration, evaporation from the plant canopy and the soil and sublimation when snow is present [8] . The rate of AET for a given environment is a function of four critical factors: soil moisture, plant type, stage of plant development, and weather conditions. In addition, the transpiration rate is influenced by soil management practices [7] [9] . McCuen [10] asserts that the most influential climatic factors on ET are temperature, relative humidity, radiation rates and wind speed.

Large AET rates in agricultural regions impact water quantity and quality [11] [12]. Accordingly, the amount of water needed for irrigation of agricultural systems is frequently calculated based on estimated AET. Underestimation of AET leads to an insufficient water supply to the crop and consequently to water stress, while overestimation of AET leads to a waste of water for irrigation. Therefore, accurate estimation of AET plays a critical role in the assessment, planning and management of water resources in general [11] . Several methods are used to estimate AET based on mass transfer, energy budget, water budget, soil moisture budget, ground-water fluctuations or meteorological variables. Each method has advantages and disadvantages; moreover, none is applicable under all conditions because of assumptions in the method or type of required data [13]. Most of the hydrological models employ the concept of potential evapotranspiration (PET) as the driving function for the calculation of AET [7] [14] [15] . PET is defined as the amount of water that would be removed from a vegetated landscape with no restrictions other than the atmospheric demand [16] , and is often measured with evaporation pans. Also, PET can be estimated using methods based on available climate data [7] [17] . Currently, a number of methods that vary in complexity and data requirements are employed for calculating PET [2] [7] [9] . Their ability to produce consistent and meaningful PET estimates depends on their assumptions, data requirements and consideration of atmospheric factors [2] [15] . HC&S has remained in the sugarcane industry for more than 100 years and during this time the company has made numerous innovations on sugarcane technology, irrigation techniques, and crop management for the sugarcane plantation on the Island of Maui. To optimize water usage for irrigation purposes, HC&S has developed a locally calibrated PET method, which is based on pan evaporation measurements and is specific to sugarcane.

The main objective of this study was to evaluate the efficiency of the HC&S irrigation system by comparing the accuracy of AET estimation by the HC&S method against three physically-based methods: Penman-Monteith, Priestley-Taylor, and Hargreaves, as well as, to a set of historical pan evaporation data and to identify an AET method that is applicable to other crops and transferable to locations with similar characteristics.

2. DATA AND METHODS

2.1. SWAT Model

SWAT is a physically based, semi-distributed, and long-term continuous watershed scale model that runs on a daily time step to predict the impact of climate, landuse, soil type, topographic characteristics, and land management practices on hydrology, sediment, nutrients, pesticides, and bacteria in large ungauged watersheds [18] [19] . SWAT estimates PET using physically based methods that are based on historical weather variables and plant/soil available water. The methods included are Hargreaves, Priestley-Taylor, and Penman-Monteith.

Runoff is estimated by the SCS Curve Number (CN) method (SCS, 1986) or the Green & Ampt method [20] . If the Green and Ampt method is used to calculate surface runoff, the model calculates canopy storage and evaporates any readily available water present in the canopy. In contrast, the CN method lumps canopy interception, surface storage and infiltration in its initial abstraction term. After calculating runoff, SWAT calculates the amount of transpiration under ideal conditions as a function of available water in the soil. Sublimation will occur if snow is present; otherwise, the actual amount of water evaporated from the soil is calculated as a function of soil depth, water content and the above-ground biomass.

SWAT simulates the water balance according to Equation (1), which includes AET, canopy interception, plant transpiration and soil evaporation, surface runoff, and vertical water movement in the unsaturated soil zone to the ground water:

$$\mathrm{SW}_i = \mathrm{SW}_0 + \left(\mathrm{R}_i + \mathrm{I}_i\right) - \left(\mathrm{Q}_i + \mathrm{AET}_i + \mathrm{PER}_i\right) \tag{1}$$

where, SW_i is the final soil water content (mm) in day i, SW_0 is the initial soil water content (mm), i is the time (days), R_i is the amount of precipitation (mm·day^{-1}), I_i is the amount of daily irrigation (mm·day^{-1}), Q_i is the amount of surface runoff (mm·day^{-1}), AET_i is the amount of actual evapotranspiration (mm·day^{-1}), and PER_i is the amount of water percolating through the soil profile (mm·day^{-1}).

2.2. PET Methods

The Penman-Monteith method [21] (Equation (2)) is a theoretically based approach that incorporates energy and aerodynamic considerations [22] . It includes a measure of the resistance to the diffusion of water vapor into the

Penman's equation, which is a radiation-aerodynamic combination equation to predict evaporation from open water, bare soil, and grass [23] :

$$\lambda E = \frac{\Delta \times (H_{net} - G) + \rho_{air} \times c_p \times (e_z^0 - e_z)/r_a}{\Delta + \gamma \times \left(1 + \frac{r_c}{r_a}\right)} \quad (2)$$

where: l is the latent heat flux density ($MJ \cdot m^{-2} kg^{-1}$), E is the depth rate evaporation ($mm \cdot day^{-1}$), Δ is the slope of the saturation vapor pressure-temperature curve ($kPa°C^{-1}$), H_{net} is the net radiation ($MJ \cdot m^{-2} d^{-1}$), G is the heat flux density to the ground ($MJ \cdot m^{-2} d^{-1}$), ρ_{air} is the air density ($kg \cdot m^{-3}$), c_p is the specific heat constant pressure ($kPa \cdot kg^{-1} °C^{-1}$), e_z^0 is the saturation vapor pressure of air at height z (kPa), e_z is the water vapor pressure of air at height z (kPa), γ is the psychometric constant ($kPa°C^{-1}$), r_c is the plant canopy resistance ($s \cdot m^{-1}$), and r_a is the diffusion resistance of the air layer (aerodynamic resistance) ($s \cdot m^{-1}$).

The Priestley-Taylor method [24] (Equation (3)) is an energy-based approach. According to Jensen et al. [16] under wet conditions, evaporation from surfaces could be estimated using a simplified version of the Penman equation, in which the aerodynamic was deleted and the energy term is assumed to be a constant fraction $\alpha_0 = 1.28$.

$$\lambda E_0 = \alpha_0 \times \frac{\Delta}{\Delta + \gamma} \times (H_{net} - G) \quad (3)$$

where: l is the latent heat of vaporization ($MJ \cdot kg^{-1}$), E_0 is the PET ($mm \cdot d^{-1}$), α_0 is a coefficient, Δ is the slope of the saturation vapor pressure-temperature curve ($kPa°C^{-1}$), γ is the psychometric constant ($kPa°C^{-1}$), H_{net} is the net radiation ($MJ \cdot m^{-2} d^{-1}$), and G is the heat flux density to the ground ($MJ \cdot m^{-2} d^{-1}$). In semiarid conditions this equation will underestimate PET.

The Hargreaves method [25] (Equation (4)) was originally derived from eight years of cool season Alta fescue grass lysimeter data from Davis, CA. According to Hargreaves and Allen [23] the main advantage of the Hargreaves approach to the other methods is the reduced data requirement.

$$\lambda E_0 = 0.002 \times H_0 \times (T_{max} - T_{min})^{0.5} \times (\overline{T}_{av} + 17.8) \quad (4)$$

where: l is the latent heat of vaporization ($MJ \cdot kg^{-1}$), E_0 is the PET ($mm \cdot d^{-1}$), H_0 is the extraterrestrial radiation ($MJ \cdot m^{-2} d^{-1}$), T_{max} is maximum air temperature for a given day (°C), T_{min} is minimum air temperature for a given day (°C), and T_{av} is the mean air temperature for a given day (°C).

The HC&S method (Equation (5)) is specific for sugarcane and also site-specific to the central isthmus of Maui. It was developed to take into account that evapotranspiration is greatly modified by the marine surroundings and the topographic characteristics of the Island of Maui. The HC&S method to calculate PET is an empirical relationship that relates extensive pan evaporation data collected in the central isthmus of Maui [26] and locally measured meteorological parameters, such as air temperature, humidity, solar radiation and wind.

This method calculates PET by including a coefficient that multiplies the saturation vapor pressure, average actual vapor pressure, the slope of the saturation vapor pressure, a weighting function for temperature, net radiation expressed as equivalent evaporation and wind function.

$$E_0 = C \times W \times H_{net} + C \times (1 - W) \times (e_z^0 - e_z) \times V_f \tag{5}$$

where E_0 is the PET ($mm \cdot d^{-1}$), C is a correction coefficient, H_{net} is the net radiation expressed in equivalent evaporation ($mm \cdot d^{-1}$), W is a weighting function for temperature, e_z^0 is the saturation vapor pressure of air at average temperature (kPa), e_z is the actual water vapor pressure of the air (kPa), and V_f is a wind function.

The correction coefficient (C) is calculated with a polynomial equation that includes eight coefficients and the ratio of average diurnal and nocturnal wind speed, solar radiation expressed in equivalent evaporation, and the mean wind speed. PET is multiplied by a crop coefficient (K_c) to calculate AET. To account for seasonal differences in crop growth and AET, two different crop coefficient curves are used depending on the time of planting.

3. STUDY AREA

The study area is located on the Island of Maui, Hawaii (20˚54' latitude and 156˚26' longitude) (Figure 1) and consists of sugarcane fields (184.3 km^2) that lie on elevations between sea level and 332 m. The climate of Maui is controlled primarily by topography and the position of the North Pacific anticyclone and other migratory weather systems relative to the island [11] . The climate is characterized by mild temperatures, cool and persistent trade winds, a rainy winter season from October through April, and a dry summer season from May through September [8] [27] . Solar energy and length of day are relatively uniform throughout the year and the surrounding ocean provides moist air and keeps temperatures fairly constant without extremes throughout the year [27] . Mean annual precipitation on Maui Island ranges from 7000 mm to 400 mm or less [28] . According to NOAA [29] , based on a 30-year dataset from the Kahului Station, Maui's Central Valley, where the HC&S Co. is located, receives an average rainfall of 453 mm per year and has an average daily temperature of 24.3˚C (ranges from 19.7˚C to 29.0˚C) and an average wind velocity of 5.7 m·s^{-1}. Relative humidity ranges from 53% to 81%.

Two fields were selected for this analysis: 905 and 415. They are located in contrasting micro-climates and landscapes. Field 905 is located at a lower elevation (31.0 m a.s.l.) with a slope of 4.3%. The soil texture is classified as silt-loam and the soils are in hydrologic group B. In contrast, Field 415 is located at a higher elevation (195.4 m a.s.l.) with a slope of 6.0%. The soil texture is classified as silty-clay-loam and the soil belongs to hydrologic group C [30] . Solar radiation (Figure 2(a)), wind velocity (Figure 2(b)), average temperature (Figure 2(c)) and relative humidity (Figure 2(d)) are plotted for both fields. There are no significant differences between fields 415 and 905 with regard to average temperature (23.7˚C and 23.6˚C), relative humidity (0.71 and 0.68), and solar radiation (18.9 and 20.0 MJ·m^{-2}). However, wind velocity shows significant differences between both fields (2.5 and 5.8 m·s^{-1}), with higher speeds found in Field 905. Relative differences for average temperature, relative humidity and solar radiation are: 2.8%, 5.9% and 2.2% respectively, while the relative difference for wind velocity is 54.9%. Temperatures and humidity are higher at Field 415 while solar radiation and wind speed are higher at Field 905. On Field 415, the sugarcane was planted in winter, while on Field 905 the crop was established in summer.

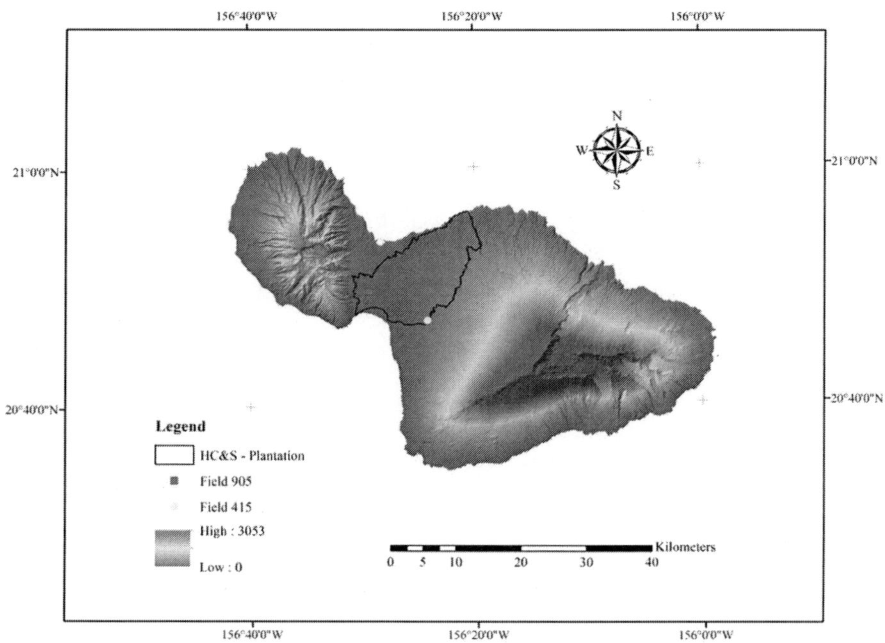

Figure 1. Location of the HC&S plantation on the Island of Maui, Hawaii.

3.1. Input Data and Model Setup

The Digital Elevation Model (DEM) with a resolution of 10 m (1/3 arc-second) was acquired from the National Elevation Dataset (NED) [31] (available at http://viewer.nationalmap.gov/viewer/). Detailed soil map and properties for the study area were obtained from the Natural Resources Conservation Service (NRCS)—Soil Survey Geographic (SSURGO) Database (Soil Survey Staff NRCS-USDA, 2011) available at http://soildatamart.nrcs.usda.gov. The soils in the HC&S plantation area are classified as Mollisols (46.3%), Andisols (15.4%) and Aridisols (15.1%), Oxisols (10.4%), Entisols (6.0%), Inceptisols (4.2%) and Ultisols (2.3%). The landuse map and related information were obtained from the United States Geological Service (USGS) Land Cover Institute (LCI) [32] available at the USGS-Seamless Data Warehouse (http://seamless.usgs.gov/) with a resolution of 30 m (1 arc-second). The agricultural land was characterized as a 2-year sugarcane crop. The ArcGIS Interface for SWAT2012 was used to delineate the boundaries of 701 sugarcane plots, defining each plot as an individual Hydrologic Response Unit (HRUs) with sizes ranging from 1.4 to 166.0 ha.

The model uses a 12-year dataset of daily climatic records from 39 rain gauges and 12 weather stations maintained by HC&S. The climatic variables included are precipitation, minimum and maximum temperature, relative humidity, solar radiation and wind speed. Missing values in the dataset were simulated by the model's builtin weather generator based on the weather statistics of the study area. HC&S applies water to the sugarcane fields through a drip irrigation system. The company maintains records of the applied volumes of water and the date of irrigation for each individual field. The SWAT model source code was modified to automatically read a database of dates and measured volumes of water applied to each field for irrigation purposes. For this study, historical daily irrigation data from 2003 to 2012 were acquired from the HC&S archive. Based on data availability, the simulation period was set up for 10 years (2003-2012) with the first year of simulation set as a spinup period. However, the analysis of model output was performed for the crop period that comprises the years 2007 through 2009.

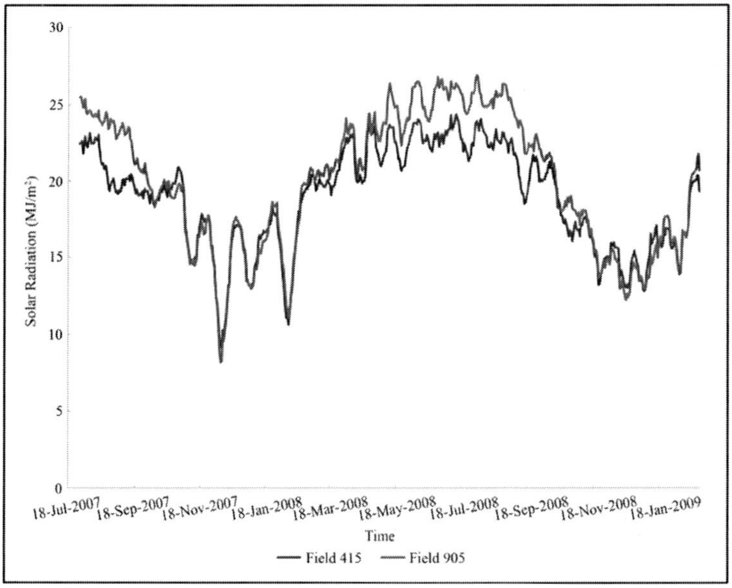

Solar Radiation (MJ/m²)
30
25
20
15
10
5
0
18-Jul-2007
18-Sep-2007
18-Nov-2007
18-Jan-2008
18-Mar-2008
18-May-2008
18-Jul-2008
18-Sep-2008
18-Nov-2008
18-Jan-2009
Time
Field 415
Field 905

(a)

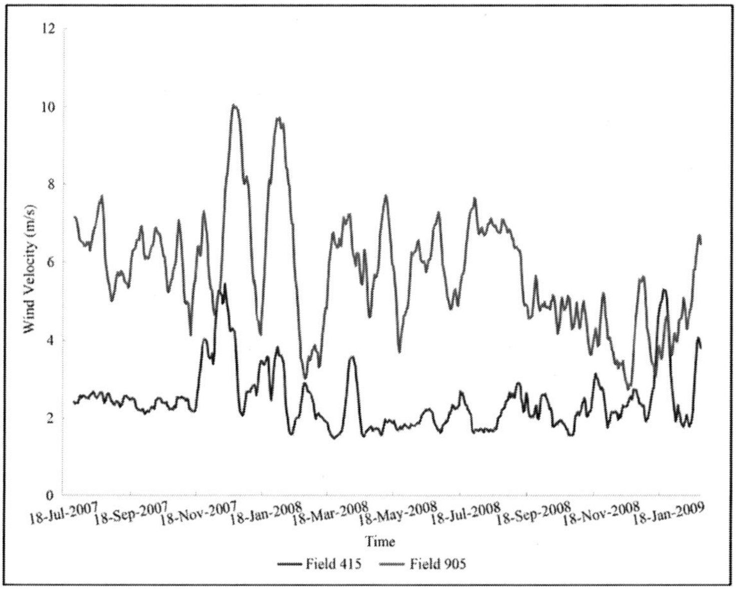

Wind Velocity (m/s)
12
10
8
6
4
2
0
18-Jul-2007
18-Sep-2007
18-Nov-2007
18-Jan-2008
18-Mar-2008
18-May-2008
18-Jul-2008
18-Sep-2008
18-Nov-2008
18-Jan-2009
Time
Field 415
Field 905

(b)

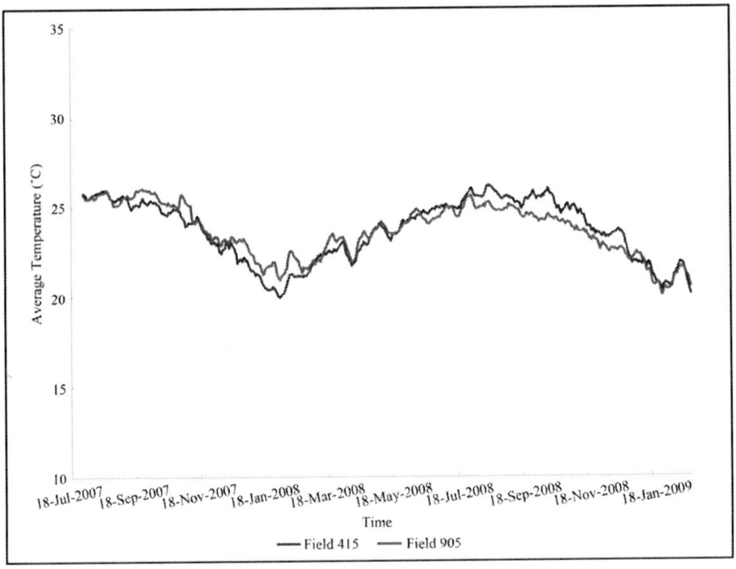

(c)

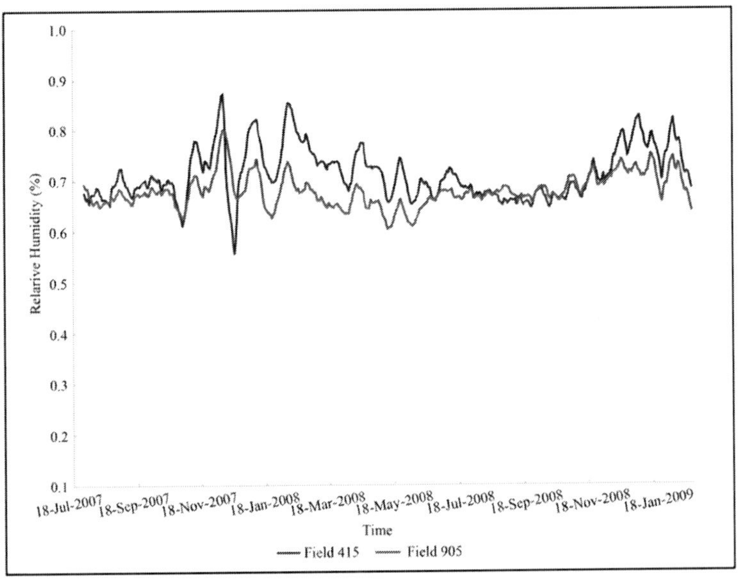

(d)

Figure 2. Measured climatic variables for the crop period at fields 415 and 905: (a) Solar radiation ($MJ \cdot m^{-2}$); (b) Wind velocity ($m \cdot s^{-1}$); (c) Average temperature (˚C); and (d) Relative humidity (%).

The water balance analysis was done evaluating surface runoff, percolation, lateral flow, precipitation, irrigation and available soil water. In addition, the relative difference between the predicted average annual PET and measured pan evaporation was calculated for the four PET methods included in this study. Measurements of pan evaporation in Hawaii began in 1894 and include intermittent observations for over 200 sites [26] . Most sites had standard, galvanized Class A pans on the surface, but many were later replaced with stainless steel pans to avoid corrosion and set on 1.52 m (equivalent to 5 ft) high platforms. The pan evaporation data used in this study ware corrected to account for differences due to height and pan material. A total of 12 galvanized Class A pans provided a data set with accumulated monthly and annual evaporation for the interval between 1963 and 1983. There were no Pan A measurements at the fields evaluated in this study; therefore, the closest available information to each field was used for comparison.

3.2. Crop Yield Calibration

The water balance in agricultural watersheds is influenced by crop production [33] . Crop dry matter production in Hawaii depends on the amount of water available from rainfall and irrigation [34] . The yield of a given crop can generally be described as a function of cumulative AET [35] . The AET has been shown to be directly related to dry biomass production when factors such as fertility, sunshine, temperature, and soil moisture are not limiting [36] . The following SWAT parameters were used for the purpose of manually calibrating crop yield: the soil evaporation compensation coefficient (ESCO), the plant uptake compensation factor (EPCO) and the biomassenergy ratio (BIO_E) ($kg \cdot ha^{-1}/MJ \cdot m^{-2}$). ESCO adjusts the depth distribution of soil evaporation to meet soil evaporative demand and varies between 0.01 and 1.0. As the value of ESCO is reduced, the model is able to evaporate more water from deeper layers in the soil profile. EPCO compensates water between soil horizons to meet the potential water uptake and ranges from 0.01 to 1.0. As EPCO approaches 1.0, the model allows more of the water uptake demand to be met by lower layers in the soil. BIO_E is the amount of dry biomass produced per unit intercepted solar radiation in ambient CO_2 and varies between 10 and 90. The greater BIO_E, the greater the potential increase in total plant biomass on a given day. An important parameter governing crop growth in SWAT is the leaf area index (LAI), which controls the amount of light intercepted by the plant and converted to biomass. Also, LAI is used to calculate AET. When LAI is larger than 3, SWAT assumes that AET equals PET. To calibrate the LAI curve calculated by SWAT, it was compared to the crop coefficient curve used by the HC&S method. To evaluate crop yields simulated by SWAT, they were compared to field measurements of crop yield obtained between 2003 and 2012.

4. RESULTS

4.1. Calibration of Sugarcane Yield

While the default values for the parameters ESCO and EPCO of 0.95 and 1.0 and BIO_E ($kg \cdot ha^{-1}/MJ \cdot m^{-2}$) of 25 were used to initialize the model; the calibrated values for the same parameters were 0.5, 0.98, and 22.5, respectively. By calibrating BIO_E, the leaf area index curve break point (LAI = 3) was adjusted to the break point of the crop coefficient (K_c = 1) used by HC&S for both winter (Field 415) and summer (Field 905) planting (Figure 3(a) and Figure 3(b)). Simulated average crop yields for the 2003-2012 period of time are 71.7 $t \cdot ha^{-1}$ (±11.3 $t \cdot ha^{-1}$) and 77.7 $t \cdot ha^{-1}$ (±12.0), for fields 415 and 905 respectively. Field measurements of crop yield for fields 415 and 905 provide an average value of 71.3 $t \cdot ha^{-1}$ (±7.7 $t \cdot ha^{-1}$) and 77.3 $t \cdot ha^{-1}$ (±10.0 $t \cdot ha^{-1}$), respectively. Predicted crop yields were not different from measured values for both fields (relative differences lower than 1%). No other goodness-of-fit metrics were used in this study due to limitations in measured data. Successful calibration of crop yield guaranteed that AET was adequately represented by the model and varied accordingly as a function of climatic variables and crop development. In addition, properly predicted AET assured appropriate partitioning of the remaining components of the water budget (Figure 3(a), Figure 3(b)).

4.2. Potential Evapotranspiration (PET)

The PET values estimated by the four methods shared a concurrent spatial pattern and temporal trend. Calculation of PET with Hargreaves or Priestley-Taylor produced values that strongly deviate from the HC&S and Penman-Monteith methods. These differences are more obvious during the summer months (May to October) than during the winter months (November to April). A comparison of daily series of PET obtained with different methods is presented in Figure 4(a), Figure 4(b). At Field 415, predicted annual PET is 1964.3 mm using the HC&S method, 1693.6 mm using Penman-Monteith, 1656.5 mm using Hargreaves, and 1542.2 mm using Priestley-Taylor. At the nearest site, measured pan evaporation has an average of 2265.5 mm ± 193.3 mm. The relative errors of the simulated values with respect to pan measurements are 13.3% for the HC&S method, 25.2% for Penman-Monteith, 26.9% for Hargreaves, and 31.9% for Priestley-Taylor. At Field 905, PET calculated using the HC&S, Penman-Monteith, Priestley-Taylor, and Hargreaves methods is 2320 mm, 2059.5 mm, 1447.2 mm, and 1157.9 mm, respectively. Measured pan evaporation nearest to Field 905 has an average of 2466.55 mm ± 244.5 mm. The relative errors are 5.9%, 16.5% mm, 41.3%, and 53.1% for the

HC&S, Penman-Monteith, Priestley-Taylor, and Hargreaves methods, respectively.

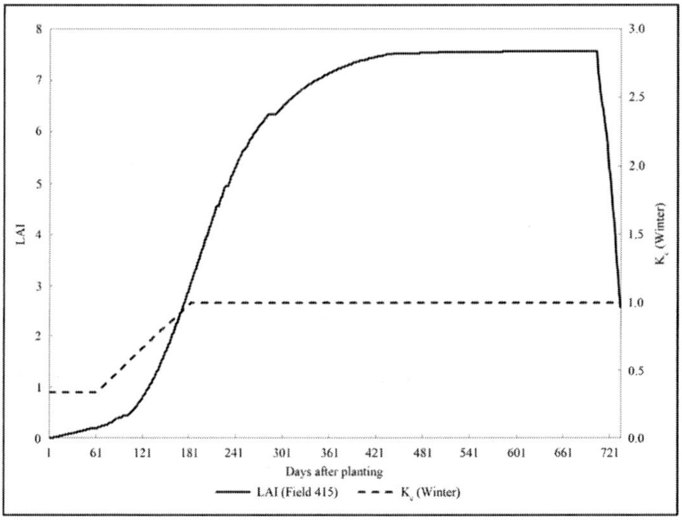

(a)

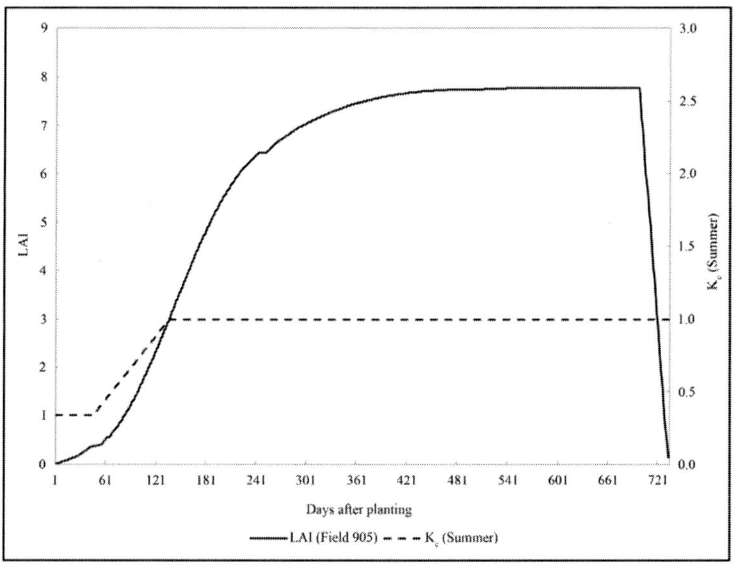

(b)

Figure 3. Leaf area index (LAI) compared to crop coefficient (K_c) for (a) the summer crop on field 415 and (b) the winter crop on field 905.

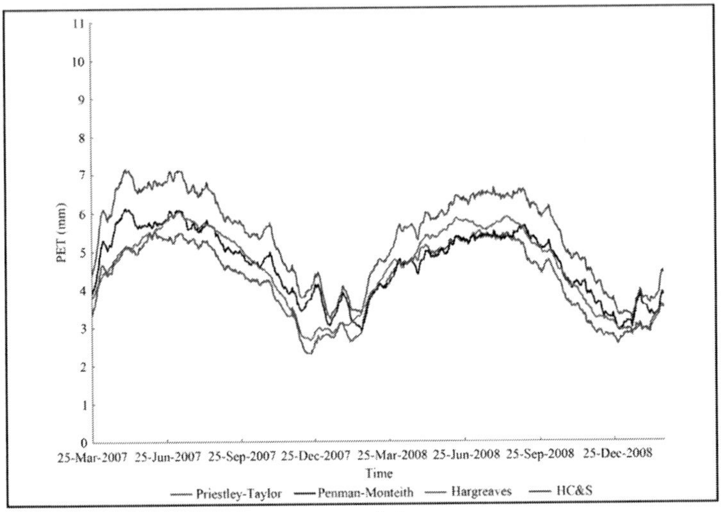

(a)

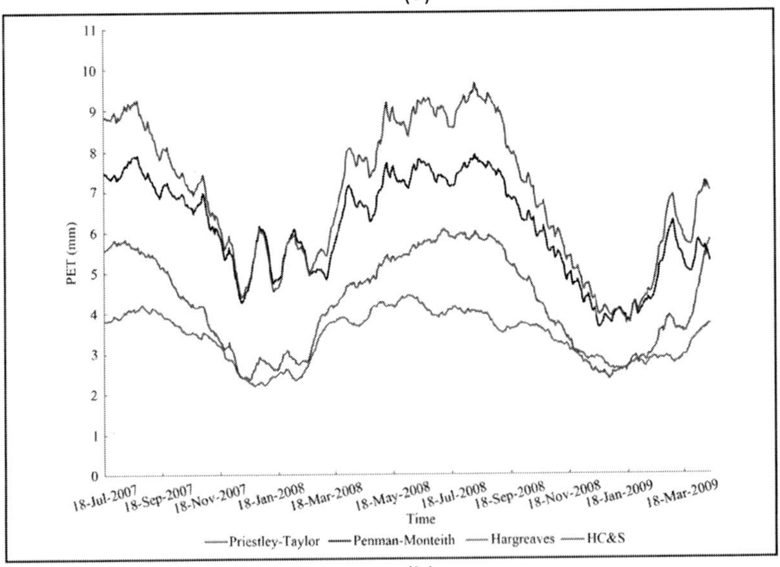

(b)

Figure 4. Comparison of PET calculated by the four tested methods at Fields 415 (a) and 905 (b).

4.3. Actual Evapotranspiration (AET)

SWAT calculates AET based on LAI or the crop coefficient method until LAI equals 3. After that, the crop is assumed to be fully developed and the model assumes that all PET is partitioned to plant water uptake and soil evaporation is assumed to be negligible. However, regardless of the crop stage, soil moisture is the limiting factor that determines the volume of water that is evaporated in a given day. This critical process is well represented in SWAT. Figure 5 shows the effect of shorter periods of low soil water contents, which reduce the AET rate to the minimum possible. A comparison of AET methods for both fields 415 and Field 905 are presented in Figure 5(a), Figure 5(b). Annual AET values predicted by SWAT range from 1081 mm to 1544 mm.

Table 1(a) and Table 1(b) show the average AET values and the corresponding standard deviations for four different crop stages. The crop cycle for sugarcane was divided into growth stages defined by the break points of the crop coefficient curves used by HC&S for winter and summer plantings. At Field 415 the highest average values of AET were found during the germination and establishment phase for all methods, while at Field 905 the highest average AET values correspond to the grand growth phase. One reason for this behavior is that both fields experienced several days of water stress during the grand growth phase, which caused AET to stop completely.

In addition, the ratio between AET estimated by the Penman-Monteith and HC&S methods were calculated for the period of time where PET is all partitioned to plant uptake and no soil evaporation occurs (grand growth phase). The ratio AET/PET can be interpreted as being equivalent to the pan coefficients, which are ratios of PET to pan evaporation for a given vegetative land cover. The calculated ratios are 1.16 ± 0.11 and 1.14 ± 0.12 for fields 415 and 905, respectively. The ratios found in this study are consistent with reported coefficients for sugarcane fields under Hawaiian conditions. USGS [37] used a sugarcane pan coefficient of 1.18 during the middle stage of growth. Reported by McMahon et al. [2] , sugarcane can have a pan coefficient as high as 1.2.

4.4. Water Balance

For hydrologic studies, the major purpose of estimating AET is to determine actual water that will be lost from the system (Figure 6). The SWAT simulated water balance for the crop period (2007-2009) is shown in Table2 The study site at Field 415 receives a total of 3149 mm (686 mm precipitation + 2463 mm irrigation). Field 905 receives a total of 2079 mm (675 mm precipitation + 2863 mm irrigation). For both fields, precipitation has a very small contribution to the water budget (27.8% and 23.4%). Most of the water supply was provided by irrigation (72.2% and 76.6%). While runoff and lateral flow explain between 2.4% to 5.1% of the water balance, AET and percolation account for 94.9% to 97.6% of the water balance calculated with the HC&S method. According to the four applied methods Priestley-Taylor, Penman-Monteith, Hargreaves, and HC&S, AET accounts for 66%, 71%, 69%, and 76% and for 67%, 79%, 58%, and 81% of the total water supply at Field 415 and Field 905, respectively. The relative contribution of the runoff is very small for both fields (0.4% to 4.3%).

Lateral flow also shows little variation among different methods (7.8% - 16.4%). The proportions of percolated water represent 26.5%, 24.4%, 32.1% and 16.9% of the water supply calculated with the Priestley-Taylor, PenmanMonteith, Hargreaves, and HC&S methods respectively.

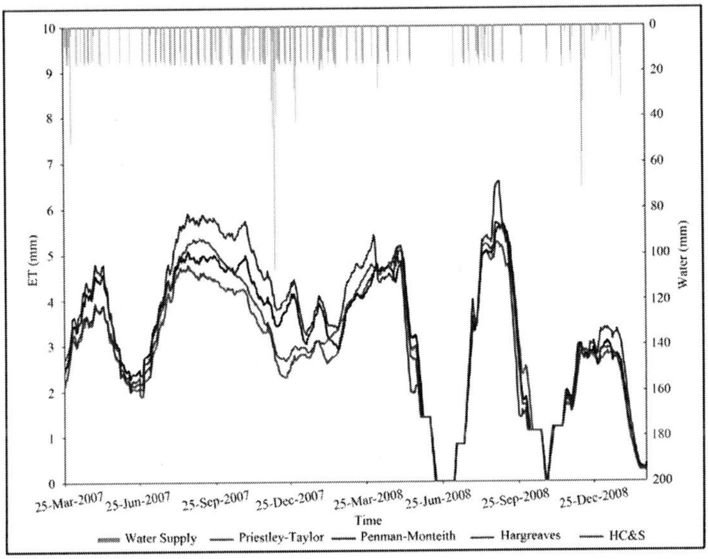

(a)

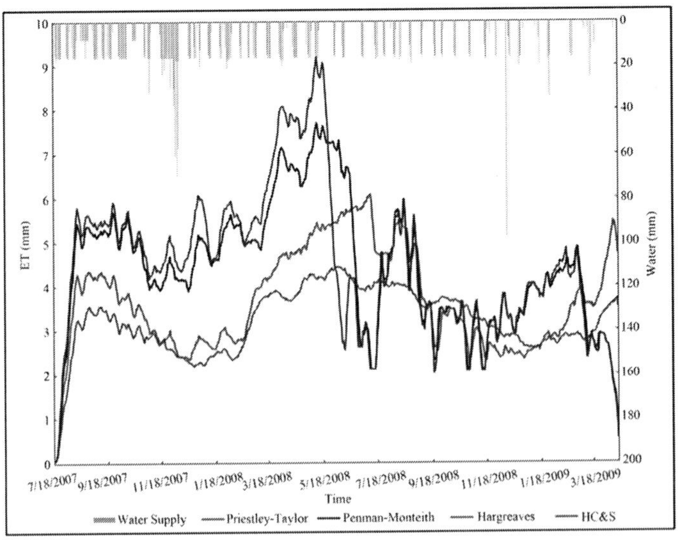

(b)

Figure 5. Comparison of actual evapotranspiration on Fields 415 (a) and 905 (b).

Table 1. Average and standard deviation of evapotranspiration by method and crop stage expressed in mm·day^{-1} for Fields 415 (a) and 905 (b).

(a)

Crop Phase	Hargreaves		Priestley-Taylor		Penman-Monteith		HC&S	
	Mean	SD	Mean	SD	Mean	SD	Mean	SD
Germination	3.6	1.3	4.1	1.5	3.6	1.2	4.3	1.9
Tillering	3.1	1.4	3.4	1.4	3.3	1.5	3.6	1.8
Grand Growth	3.2	1.8	3.5	1.9	3.4	1.9	3.8	2.3
Ripening	2.3	2.0	2.3	2.2	2.3	2.1	2.4	2.5
All Crop Period	3.0	1.8	3.2	2.0	3.1	1.9	3.4	2.4

(b)

Crop Phase	Hargreaves		Priestley-Taylor		Penman-Monteith		HC&S	
	Mean	SD	Mean	SD	Mean	SD	Mean	SD
Germination	3.9	1.4	4.9	1.8	3.1	1.1	5.2	2.0
Tillering	3.1	1.0	4.7	1.4	2.7	0.7	5.1	1.6
Grand Growth	4.5	1.8	5.1	3.0	3.8	0.6	5.1	3.8
Ripening	3.2	1.5	3.2	2.3	3.0	0.6	3.2	2.6
All Crop Period	3.8	1.6	4.4	2.6	3.3	0.8	4.5	3.1

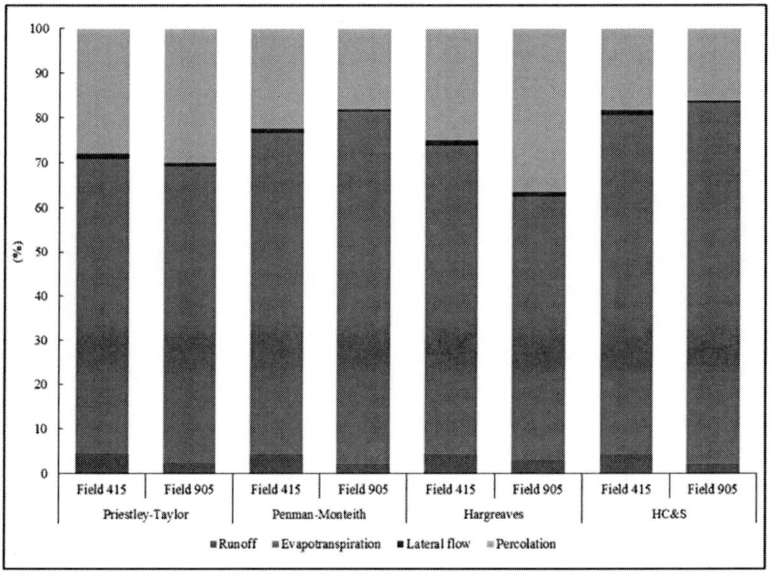

Figure 6. Water balance for Fields 415 and 905 expressed in percent of water supply.

Table 2. Water balance for Fields 415 (a) and 905 (b) expressed in mm of water. Results of the 2-year crop period.

(a)

Component	Priestley-Taylor	Penman-Monteith	Hargreaves	HC&S
Total Water Inflow	**3166.9**	**3162.4**	**3165.1**	**3147.6**
Soil Available Water Content	17.9	13.4	16.1	−1.4
Total Water Supply	3149.0	3149.0	3149.0	3149.0
Precipitation	685.7	685.7	685.7	685.7
Irrigation	2463.2	2463.2	2463.2	2463.2
Total Water Outflow	**3128.5**	**3132.8**	**3130.1**	**3145.0**
Runoff	135.7	131.8	132.2	129.4
Evapotranspiration	2079.5	2264.7	2175.8	2403.4
Lateral Flow	35.7	32.8	34.2	30.8
Percolation	877.7	703.5	787.8	581.5

(b)

Component	Priestley-Taylor	Penman-Monteith	Hargreaves	HC&S
Total Water Inflow	**3567.5**	**3567.9**	**3644.0**	**3567.6**
Soil Available Water Content	0.0	0.4	76.5	0.0
Total Water Supply	3567.6	3567.6	3567.6	3567.6
Precipitation	675.8	675.8	675.8	675.8
Irrigation	2891.7	2891.7	2891.7	2891.7
Total Water Outflow	**3561.5**	**3561.1**	**3485.3**	**3561.2**
Runoff	75.1	67.5	95.8	67.4
Evapotranspiration	2389.7	2828.8	2083.5	2896.9
Lateral Flow	25.0	17.6	28.5	16.4
Percolation	1071.7	647.2	1277.5	580.4

ence evapotranspiration in irrigated regions of Maui under windy conditions. In that regard, Hargreaves and Allen [23] stated that daily estimates by the Hargreaves equation are subject to error because of the influence of the temperature range caused by large variations in wind speed. Both Hargreaves and Priestley-Taylor methods produced predictions that tended to be smaller than those predicted by the Penman-Monteith and HC&S methods. Dingman [8] concluded that temperature is not a reliable indicator of sunlight and generally is a poor predictor of pan evaporation because air temperature in Hawaii is greatly modified by the marine surroundings. Under the conditions of the study site, whenever soil water content is not the limiting factor, differences in wind velocity account for most of the variation in PET calculated with different methods. In general, PET methods, that require wind and especially humidity data, such as Penman-Monteith and HC&S, will simulate windy and humid conditions more accurately [14] [38] . The observed trends of

AET as a function of crop growth stage failed to match the trends reported in other publications in which AET rises from germination phase through tillering and reaches the maximum value at the grand growth phase to then decline during the ripening phase. For instance, Freire [39] reported AET values of 3.0 $mm \cdot day^{-1}$ for the germination phase, 3.8 $mm \cdot day^{-1}$ for tillering phase, 5.1 $mm \cdot day^{-1}$ for the grand growth period, and 3.1 $mm \cdot day^{-1}$ for the ripening phase. One reason for this result could be water stress. Sugarcane at both fields had to endure a lack of rainfall and irrigation for several days at late states of the crop development. As reported, AET in sugarcane fields is highly variable largely due to local climatic conditions, hydrologic settings, and the type of production systems. However, predicted AET made by SWAT for the entire crop cycle with Penman-Monteith and HC&S methods in two different field conditions compared well to reported measured data of AET that ranges between 2.33 - 5.70 $mm \cdot day^{-1}$ [39] -[42] .

Using the HC&S method, the proportion of percolated water (16.3% - 18.5%) compares reasonable well to 19% and 34% of recharge found as a proportion of the total water input by Izuka et al. [43] and Shade [44] . In addition, AET proportions calculated with the HC&S method (76.4% - 81.3%) are also supported by the comparisons made with measured Pan A evaporation datasets. Accordingly, it can be assumed that SWAT properly represents the system under analysis when using the HC&S method to account for AET. HC&S managers use their own modified ET equation to plan irrigation. In both fields (415 and 905), the amounts of irrigation (2463.2 mm and 2896.9 mm), i.e. the water supply, and AET (2403.4 mm and 2896.9 mm), i.e. the demand, are similar.

6. CONCLUSIONS

Considering the important role that evapotranspiration plays in the water cycle, the appropriate selection of the PET method to estimate AET is of great importance. Commonly, the PET method is selected based on the availability of climatic datasets to feed the models. The lack of climatic variables limits the use of physically sound PET approaches. However, the selection of the PET method should rather be taken based on site characteristics; the use of a PET method that takes into account the dominant climatic variables is highly recommended. By ignoring the dominant weather factor, predicted PET can be meaningless and likely not representative of the environmental conditions that need to be assessed. Based on the results of this study, the HC&S method appropriately represents the specific climatic and environmental conditions of the central valley in Maui, in which HC&S sugarcane plantations are located. However because the HC&S method was developed for very specific conditions, the Penman-Monteith method is recommended for other parts of the Hawaiian Islands, where climatic conditions are dominated by wind characteristics. In order to represent local conditions, PET during the grand growth phase of sugarcane should be calculated multiplying the Penman-Monteith PET by a coefficient that ranges between 1.14 and 1.16.

The present study assessed the influences of PET methods on the water balance of sugarcane cropping system in the Hawaiian Island of Maui. The results suggests that the proportions of ET, percolated water, lateral flow and runoff compare reasonably well with results reported in other studies performed on the Island of Maui. Accordingly, the HC&S Company allocates water for irrigation based on an appropriate estimation of AET. However, additional work should be done to analyze the entire system that includes water collection from the ditch system (including bypass water), analysis of ground water recharge and pumping system and the water budget for the entire island.

ACKNOWLEDGEMENTS

This study was funded by the US Department of Navy, Office of Naval Research & USDA-ARS Bioenergy Project in Hawaii. The authors are grateful to Mae Nakahata from HC&S for providing research access and assistance and to the anonymous reviewers for their valuable suggestions to improve the paper.

REFERENCES

1. Grubert, E. (2011) Freshwater on the Island of Maui: System Interactions, Supply, and Demand. Master's Thesis, Environmental Water Resource Engineering, University of Texas at Austin, Austin.
2. McMahon, T.A., Peel, M.C., Lowe, L., Srikanthan, R. and McVicar, T.R. (2012) Estimating Actual, Potential, Reference Crop and Pan Evaporation Using Standard Meteorological Data: A Pragmatic Synthesis. Hydrology and Earth System Sciences, 9, 11829-11910.
3. Lloyd, B.R., Davidson, J.R. and Hogg, H.C. (1972) Estimating the Productivity of Irrigation Water for Sugarcane Production in Hawaii. Economic Research Service Technical Report No. 56, USDA—Natural Resource Economics Division, 72.
4. National Agricultural Statistics Service (NASS) (2013) National Statistics for Sugarcane. http://www.nass.usda.gov
5. Baumgartner, A. and Reichel, E. (1975) The World Water Balance. R. Oldenbourg-Verlag, Munchen, Wien, 1-179.
6. Irmak, S., Howell, T.A., Allen, R.G., Payero, J.O. and Martin, D.L. (2005) Standardized ASCE Penman-Monteith: Impact of Sum-Of-Hourly vs.

24-Hour Timestep Computations at Reference Weather Station Sites. Transactions of the ASAE, 48, 1063-1077.

7. Wang, X., Melesse, A.M. and Yang, W. (2006) Influences of Potential Evapotranspiration Estimation Methods on Swat's Hydrologic Simulation in a Northwestern Minnesota Watershed. Transactions of the ASABE, 49, 1755-1772.
8. Dingman, S.L. (1994) Physical Hydrology. Macmillan College Publishing Company, New York, 575.
9. Allen, R.G., Pereira, L.S., Raes, D. and Smith, M. (1998) Crop Evapotranspiration: Guidelines for Computing Crop Water Requirements. FAO Irrigation and Drainage Paper 566, UN-FAO, Rome.
10. McCuen, R.H. (1998) Hydrologic Analysis and Design. 2nd Edition, Prentice-Hall, Upper Saddle River.
11. Giambelluca, T.W. and Nullet, D. (1992) Evaporation at High Elevations in Hawaii. Journal of Hydrology, 136, 219-235.
12. Farahani, H.J., Howell, T.A., Shuttleworth, W.J. and Bausch, W.C. (2007) Evapotranspiration: Progress in Measurement and Modeling in Agriculture. Transaction of the ASABE, 50, 1627-1638.
13. Shih, S.F. (1987) Using Crop Yield and Evapotranspiration Relations for Regional Water Requirement Estimation. Water Resources Bulletin—American Water Resources Association, 23, 435-442.
14. Gracel, B. and Quickl, B. (1998) A Comparison of Methods for the Calculation of Potential Evapotranspiration under the Windy Semi-Arid Conditions of Southern Alberta. Canadian Research Hydrology, 13, 9-19.
15. Blumenstock, D.I. and Price, S. (1967) Climate of Hawaii. In: Climates of the States, No. 60-51, Climatography of the United States, US Department of Commerce, Washington DC.
16. Jensen, M.E., Burman, R.D. and Allen, R.G. (1990) Evapotranspiration and Irrigation Water Requirements. American Society of Civil Engineers Manual and Reports on Engineering Practice No. 70, New York, 360.
17. Bezuidenhout, C.N., Lecler, N.L., Gers, C. and Lyne, P.W.L. (2006) Regional Based Estimates of Water Use for Commercial Sugarcane in South Africa. Water South Africa, 32, 219-222.
18. Arnold, J.G. and Fohrer, N. (2005) SWAT2000: Current Capabilities and Research Opportunities in Applied Watershed Modeling. Hydrolocical Processes, 19, 563-572.

19. Gassman, P.W., Reyes, M., Green, C.H. and Arnold, J.G. (2007) The Soil and Water Assessment Tool: Historical Development, Applications, and Future Directions. Transactions of the ASABE, 50, 1211-1250.
20. Green, W.H. and Ampt, G.A. (1911) Studies on Soil Physics, Part 1, the Flow of Air and Water through Soils. Journal of Agricultural Science, 4, 11-24.
21. Monteith, J.L. (1965) Evaporation and the Environment. 19th Symposia of the Society for Experimental Biology, 19, 205-234.
22. Sumner, D.M. and Jacobs, J.M. (2005) Utility of Penman-Monteith, Priestley-Taylor, Reference Evapotranspiration, and Pan Evaporation Methods to Estimate Pasture Evapotranspiration. Journal of Hydrology, 308, 81-104.
23. Hargreaves, G.H. (1975) Moisture Availability and Crop Production. Transactions of the ASAE, 18, 980-984.
24. Priestley, C.H.B. and Taylor, R.J. (1972) On the Assessment of Surface Heat Flux and Evaporation Using Large-Scale Parameters. Monthly Weather Review, 100, 81-82.
25. Hargreaves, G.H. and Allen, R.G. (2003) History and Evaluation of Hargreaves Evapotranspiration Equation. Journal of Irrigation and Drainage Engineering, 129, 53-63.
26. Ekern, P.C. and Chang, J.H. (1985) Pan Evaporation: State of Hawaii, 1894-1983. Hawaii. Department of Land and Natural Resources, Division of Water and Land Development, Honolulu, Rep. R74, viii + 172.
27. Sanderson, M. (1993) Prevailing Trade Winds: Weather and Climate in Hawaii. University of Hawaii Press, Honolulu, 126.
28. Mink, J.F. and Lau, L.S. (2006) Hydrology of the Hawaiian Islands. University of Hawaii Press, Honolulu.
29. NOAA (National Oceanic and Atmospheric Administration) (2013) Comparative Climatic Data for the United States through 2012.
30. Anderson, R.G. and Wang, D. (2014) Energy Budget Closure Observed in Paired Eddy Covariance Towers with Increased and Continuous Daily Turbulence. Agricultural and Forest Meteorology, 184, 204-209.
31. Gesch, D.B. (2007) The National Elevation Dataset. In: Maune, D., Ed., Digital Elevation Model Technologies and Applications: The DEM User's Manual, 2nd Edition, American Society for Photogrammetry and Remote Sensing, Bethesda, 99-118.
32. Fry, J.A., Coan, M.J., Homer, C.G., Meyer, D.K. and Wickham, J.D. (2009) Completion of the National Land Cover Database (NLCD) 1992-

2001 Land Cover Change Retrofit Product. US Geological Survey Open-File Report 2008-1379, 18.

33. Nair, S.S., King, K.W., Witter, J.D., Sohngen, B.L. and Fausey, N.R. (2011) Importance of Crop Yield in Calibrating Watershed Water Quality Simulation Tools. JAWRA: Journal of the American Water Resource Association, 47, 1285-1297.

34. Jones, C.A. (1980) A Review of Evapotranspiration Studies in Irrigated Sugarcane in Hawaii. Hawaiian Planters' Record, 59, 195-214.

35. Liu, W.Z., Hunsaker, D.J., Li, Y.S., Xie, X.Q. and Wall, G.W. (2002) Interrelations of Yield, Evapotranspiration, and Water Use Efficiency from Marginal Analysis of Water Production Functions. Agricultural Water Management, 56, 143-151.

36. Allison, F.E., Roller, E.M. and Raney, W.A. (1985) Relationship between Evapotranspiration and Yields of Crops Grown in Lysimeters Receiving Natural Rainfall. Agronomy Journal, 50, 506-511.

37. United States Geological Service (USGS) (2007) Effects of Agricultural Land-Use Changes and Rainfall on Ground-Water Recharge in Central and West Maui, Hawaii, 1926-2004. Scientific Investigations Report 2007-5103, 69.

38. Suleiman, A.A. and Hoogenboom, G. (2007) Comparison of Priestley-Taylor and FAO-56 Penman-Monteith for Daily Reference Evapotranspiration Estimation in Georgia. Journal of Irrigation and Drainage Engineering, 133, 175-182.

39. Da Silva, T.G.F.F. (2009) Analise de crescimento, interacao biosfera-atmosfera e eficiencia do uso de agua da cana-deacucar irrigada no submedio do vale do Sao Francisco. TeseDoutorado—Universidade Federal de Vicosa, 194.

40. Souza, E.F., Bernardo, S. and Carvalho, J.A. (1999) Funcao de producao da cana—Deacucaremrelacao a agua para tres cultivares, em Campos dos Goytacazes. Engenharia Agicola, 19, 28-42.

41. Inman-Bamber, N.G. and McGlinchey, M.G. (2003) Crop Coefficients and Water-Use Estimates for Sugarcane Based on Long-Term Bowen Ratio Energy Balance Measurements. Field Crops Research, 83, 125-138.

42. Santos, M.A.L. (2005) Irrigacao suplementar da cana-de-acucar (Saccharumspp.): Um modelo de analise de decisao para o Estado de Alagoas. Tese Doutorado—Escola Superior de Agricultura "Luiz de Queiroz", Piracicaba, 101.

43. Izuka, S.K., Oki, D.S. and Chen, C. (2005) Effects of Irrigation and Rainfall Reduction on Ground-Water Recharge in the Lihue Basin, Kauai, Hawaii. Scientific Investigations Report 2005-5146, US Department of the Interior, 48.

44. Shade, P.J. (1999) Water Budget of East Maui, Hawaii. Water-Resources Investigations Report 98-4159, US Department of the Interior, 36.

Chapter 6

Evapotranspiration Characteristics of a Lowland Dry Evergreen Forest in Central Cambodia Examined Using a Multilayer Model

Tatsuhiko NOBUHIRO[1*], Akira SHIMIZU[2], Katsunori TANAKA[3], Koji TAMAI[1], Naoki KABEYA[2], Eriko ITO[4], Takanori SHIMIZU[1], Makoto ARAKI[1], Sophal CHANN[5]

[1]Forest Hydrology Laboratory, Forestry and Forest Products Research Institute, Tsukuba, Japan

[2]Kyushu Research Center, Forestry and Forest Products Research Institute, Kumamoto, Japan

[3]Research Institute for Global Change, Japan Agency for Marine-Earth Science and Technology, Yokohama, Japan

[4]Hokkaido Research Center, Forestry and Forest Products Research Institute, Hokkaido, Japan

[5]Forestry Administration, Forest and Wildlife Science Research Institute, PhnomPenh, Cambodia

ABSTRACT

Development pressure has led to serious deforestation on the Indochina Peninsula. Particularly rapid deforestation has occurred in easily accessible lowland areas, and it is thus important to accumulate knowledge about these forests immediately. We measured evapotranspiration rates for a lowland dry evergreen forest in Kampong Thom Province, central Cambodia, using the energy balance Bowen ratio (EBBR) method based on meteorological data collected from a 60-m-high observation tower. Daily evapotran spiration was higher during the dry season than during the rainy season of the Asian monsoon climate. The seasonal variation in evapotranspiration generally corresponded to the seasonal difference in the vapor pressure deficit. A multilayer model was used to simulate the seasonal variation in evapotranspiration. The multilayer model also reproduced the larger evapotranspiration rate in the dry season than in the rainy season. However, observed values substantially exceeded model-calculated results during certain periods at the beginning of the dry season and in the late dry season. Moreover, during the rainy season, the model tended to overestimate evapotranspiration. The differences between these observed and simulated values may have been caused by seasonal characteristics of photosynthesis and transpiration in the lowland dry evergreen forest that were not considered in the model simulation.

Keywords: Cambodia, Energy Balance Bowen Ratio Method, Evapotranspiration, Lowland Dry Evergreen Forest, Multilayered Model

1. INTRODUCTION

The Indochina Peninsula, including Cambodia, belongs to the Asian monsoon climate zone characterized by distinct rainy and dry seasons. Pressures including increases in population, farmland development, and illegal logging have led to serious deforestation problems across the Indochina Peninsula. For example, the area covered by forest in Thailand decreased from 54% in 1960 to 25.3% in 1998 [1], and the forested area in Cambodia dropped from 73% in 1969 to 58% in 1997 [2].

Since artificial afforestation of commercial species such as teak has also been occurring in Indochina [3,4], very few forests exist in their natural state. To develop suitable forestry management practices for this region, a first urgent task is to assemble information on the various natural and artificial forest types, considering the climatic characteristics of the seasonal tropics.

Several studies have examined water cycles and resources in the Asian monsoon region, focusing on deforestation and certain regional climatic features. Using a gradient model, Pinker et al. [5] reported that evapotranspiration rates in the dry evergreen forest at the Sakaerat Environmental Research Station in Thailand was highest in June (rainy season; 3.3 mm day^{-1}) and lowest in January (dry season; 0.6 mm day^{-1}). Using the energy balance Bowen ratio (EBBR) method, Attarod et al. [6] also reported higher evapotranspiration rates during the rainy season (4.1 mm day^{-1}) than in the defoliated dry season (2.5 mm day^{-1}) for an artificial teak forest at Mae

Moh in northern Thailand. However, Tanaka et al. [7], measuring evapotranspiration by the eddy correlation method for a hill evergreen forest in the Kog-Ma watershed of northern Thailand, reported that transpiration peaks appeared during the late dry season. At Kog-Ma, the water budget, canopy rainfall interception, sap flow, and soil water content were also measured, and both soil moisture and root system development were determined to be sufficient throughout the year that suppressed transpiration did not occur during the dry season [8,9].

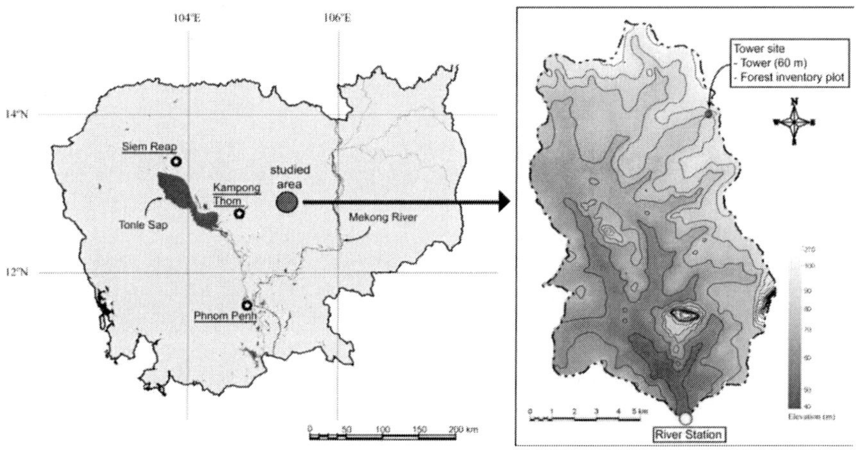

Figure 1. Location of the observation points in the O Thom I watershed.The contour interval of the topographic map is 10 m.

As noted above, in the dry season of the Indochina Peninsula, transpiration decreases in some forests, whereas brisk transpiration continues in other forests. Differences in transpiration may be caused by differences in tree species and site environments such as soil depth or root system development. Thus, it is important to accumulate information on various forest types [10,11].

Lowland dry evergreen forest was once widely distributed over the Indochina Peninsula. Currently, however, this forest type is found only in limited areas, such as Kampong Thom Province in central Cambodia. Given the strong development pressures and lack of previous research on lowland forests, the need to gather information on lowland dry evergreen forest is particularly urgent.

This paper presents the results of evapotranspiration observations using the EBBR method in a lowland dry evergreen forest located in Kampong Thom Province. The characteristics of evapotranspiration are also described based on comparison with model simulations for the study forest; the simulations were conducted by a multilayer model that was previously applied to a hill evergreen forest of the Kog-Ma experimental watershed in Thailand [7].

2. SITE DESCRIPTION AND OBSERVATIONS

2.1. Site Description

The study was conducted at the O Thom I experimental watershed in Kampong Thom Province, Cambodia. The O Thom I River is a branch of the Stung Chinit River which feeds the Tonle Sap River, a major tributary of the Mekong River in central Cambodia. The O Thom I watershed covers an area of 137 km^2, with altitude ranging between 46 and 273 m. Kabeya et al. [12] reported that annual precipitation in 2004 was around 1500 mm at the Stung Chinit River, and monthly averaged temperature ranged from 26.8 to 30.2ºC.

We constructed a 60-m-high observation tower from which meteorological elements were measured in the northeast part of the O Thom I experimental watershed (12º44'N, 105º28'E, 88 m a.s.l.; Figure 1). The topography near the tower was rather flat.

Vegetation at the experimental watershed consisted of broad-leaved evergreen trees such as dipterocarp species.

The vegetation near the meteorological observation tower was analyzed in great detail. The mean diameter at breast height and the mean tree height of overstory trees were 39.6 cm and 27.2 m, respectively. The maximum tree height was about 45 m. Although the upper canopy trees were low in density (512 stems ha^{-1}), they covered a large area of the canopy, with some secondary layer trees filling crown gaps between trees in the upper canopy [13]. Within the O Thom I watershed, the soil group was identified as acrisol [14] with a soft soil layer reaching to a depth of at least 7 m [15]. Ohnuki et al. [11] reported that the lowland dry evergreen forested area had a maximum water storage capacity of 1350 mm.

2.2. Observation Scheme

Our analysis of evapotranspiration and water balance in the O Thom I watershed parallels the study of Shimizu et al. [16], who also measured precipitation, runoff, shallow groundwater level, soil moisture, and meteorological factors in an experimental watershed of Kyusyu Island in southeastern Japan.

Table 1. Parameters used in the simulation.

Parameter	Symbol	Value	Reference or Remarks
	Physical Characteristics of an Individual Leaf		
Drag coefficient on both leaf surfaces	C_d	0.2	[24]
Bulk coefficient for sensible heat on both leaf surfaces	C_h	0.06	[26]
	Optical Characteristics of an Individual Leaf		
Transmissivity of solar radiation and PAR	τ	0.2 and 0.06	[27]
Reflectivity of solar radiation and PAR	ρ	0.3 and 0.09	[27]
	Biological Characteristics of an Individual Leaf		
Slope[a]	m	9	parameter tuning referred to Tanaka et al. [7]
Maximum rate of carboxylation at 25°C[b]	V_{cmax25}	20 μmol m^{-2} s^{-1}	parameter tuning referred to Tanaka et al. [7]
	Other Characteristics of an Individual Leaf		
Surface emissivity	ε	1.0	
Capacity of water storage on the upper side	W_{Umax}	0.1 mm LAI^{-1}	parameter tuning referred to Tanaka [25]
Capacity of water storage on the lower side	W_{Lmax}	0.1 mm LAI^{-1}	Assumed equal to W_{UMAX}
	Canopy Structure		
Number of canopy layers		50	
Canopy height	h	33 m	[13]
Leaf Area Index	LAI	3.87 - 4.49	see Tab. 2
Vertical profile of Leaf Area Density	a(z)	D-Type	a(z) is regarded as beta distribution [7,13]
Distribution of leaf inclination	$g(\alpha)$	60° (SD=18)	$g(\alpha)$ is regarded as normal distribution [7]
	Characteristics of the Soil		
Reflectivity of solar radiation and PAR	α_{soil}	0.26 and 0.06	[26]
Roughness length	z_0	0.02 m	[26]
Moisture availability	β_{soil}	0	soil evaporation is assumed to be negligible
	Other Characteristics of the Soil		
Surface emissivity	ε	1.0	[7]
Soil heat storage		0 W m^{-2}	[7]

[a]For stomatal model

[b]For net assimilation model

Pyranometers (CM3, Kipp & Zonen) were installed 60 m above the ground at the top of the observation tower to monitor solar radiation and reflected radiation. Two ventilated psychrometers (MH-020T, Eko) were installed above the forest canopy at 34 and 38 m. A net radiometer (Q7.1, REBS) and a cup anemometer (3301-5, Young) were placed at 36 m, and two heat flux plates (MF-180M, Eko) were installed 2 cm under the forest soil near the tower. Atmospheric pressure measurements (mini-TROLL, Air brown) were carried out at Kbal Domrey (12°54'N, 105°28'E) 20 km north of the tower site and recorded every hour. The leaf area index (LAI) was estimated using a LAI-2000 plant canopy analyzer (LI-COR) and hemispherical photography at the end of each month. A rain gauge (RG-2M, Onset) was installed at 60 m height on 25 April 2004. Pyrgeometers (CGR-3, Kipp & Zonen) were temporarily placed at the top of the tower, and intensive observations of upward and downward longwave radiation were carried out from 27 February to 2 March 2006. Meteorological measurements, including of soil heat flux, were taken at 10-s intervals near the tower and recorded as 10-min averages (or cumulative values) using a data logger (CR10X, Campbell). The primary meteorological observations for this study were made throughout the year in 2004.

To confirm the instrumental accuracy of the ventilated psychrometers, which has a large effect on the Bowen ratio calculation, verifications were carried out every 3 months. The psychrometers were set at the same height, and inter-variation of the equipment was checked. At the same time, dryand wet-bulb temperatures were also calibrated using an Asman psychrometer (Y-6005K, YOSHINO).

2.3. Estimation of Evapotranspiration by the Energy Balance Bowen Ratio Method

The energy balance Bowen ratio (EBBR) method, which measures the energy budget above the forest canopy from the profiles of potential temperature and water vapor pressure, was used to estimate evapotranspiration as described by Hattori [17] and Kaimal and Finnigan [18]. The method assumes that the diffusion coefficients of the latent and sensible heat fluxes are equal at the atmospheric surface layer. Therefore, the Bowen ratio (β) can be used to estimate latent heat flux as follows:

$$\beta = \frac{H}{\lambda E}, \tag{1}$$

$$\lambda E = \frac{Rn - G}{1 + \beta}, \tag{2}$$

where Rn is the net radiation (including net shortwave and longwave radiation), H is the sensible heat flux, λE is the latent heat flux, and G is the ground heat flux. The Bowen ratio can then be calculated by measuring the dryand wet-bulb temperatures at the two levels and incorporating the psychrometric constant and the gradient of the saturation vapor pressure curve as follows:

$$\beta = \frac{\gamma(TD_1 - TD_2)}{(\delta + \gamma)(TW_1 - TW_2) - \gamma(TD_1 - TD_2)}, \tag{3}$$

where γ is the psychrometric constant, δ is the gradient of the saturated vapor pressure curve at the average value of the wet-bulb temperatures, and $TD_1 - TD_2$ and $TW_1 - TW_2$ are the differences between dryand between wet-bulb temperatures at the two heights (34 and 38 m), respectively. All data required to solve these equations were collected using the instruments mentioned above.

When applying the EBBR method, atmospheric stability is preferable in a nearly neutral condition [19]. Tamai et al. [20] reported that the degree of atmospheric stability in O Thom I experimental watershed is generally suitable for tower flux measurements, including by the EBBR method. Moreover, as described above, we paid careful attention to maintaining the instrumental accuracy of the psychrometers; therefore, we expect that our observational results obtained by the EBBR method are highly reliable.

3. MULTILAYER MODEL FOR EVAPOTRANSPIRATION

The multilayer model that had been previously applied to the hill evergreen forest at Kog-Ma experimental watershed [7,9] was applied to the lowland dry evergreen forest at O Thom I experimental watershed. In this chapter, we explain the required input data and the parameters used to apply the multilayer model to our experimental watershed.

3.1. Model Structure

The multilayer model consists of distinct physical processes and contains five sub-models that incorporate the following: 1) Reynolds stress and temperature, CO_2, and H_2O exchanges on leaves; 2) radiation transfer within the forest canopy; 3) atmospheric diffusion within and above the forest canopy; 4) energy balance on leaves and the ground surface; and 5) rainfall interception and water budget on the leaf surface.

The details of each sub-model are described below.

1) The Farquhar-type photosynthesis model [21] and Ball's stomatal conductance model [22] are used for the CO_2 and H_2O exchanges of leaves.

2) The vertical profiles of leaf area density a(z) and the distribution of the leaf inclination angle g(α) are used to determine radiation transfer.

3) A second-order closure sub-model [23] is used to simulate atmospheric transfer within and above the forest canopy. This sub-model predicts the effect of a(z) on atmospheric diffusion from the soil surface to twice the canopy height [24].

4) For the energy balance sub-model, the heat storage in the leaves and the energy stored by photosynthesis are assumed to be negligible, and the radiation energy in each divided layer is estimated from the transmissivity and reflectivity of shortwave and longwave radiation apportioned into sensible and latent fluxes. The energy balance on the ground surface is estimated from the input radiation energy from the lowest layer and the soil heat flux [25].

5) The rainfall interception sub-model assumes that rainfall wets only the upper sides of leaves, whereas condensation wets both the upper and lower sides of leaves, and transpiration and CO_2 exchange occur only at dry lower-leaf stomata. Moreover, this sub-model captures the effects of a(z) and g(α) on rainfall interception, which are important during the rainy season.

The multilayer model also assumes that wind velocity, air and leaf temperatures, humidity, and CO_2 concentration are horizontally uniform, although it does consider the difference in photosynthetically active radiation (PAR) in sunlit and shaded areas [7].

3.2. Model Input

The data input to the multilayer model are solar elevation (SE), downward shortwave radiation (S↓), downward longwave radiation (L↓), horizontal wind velocity (u), air temperature (T), vapor pressure (e), CO_2 partial pressure (c), and rainfall over the forest (P). The outputs are upward shortwave radiation (S↑), upward longwave radiation (L↑), net radiation (Rn), sensible heat flux (H), latent heat flux (λE), CO_2 flux (Fc), and their respective vertical profiles. The multilayer model separates the latent heat flux into transpiration and evaporation components. Canopy structure, the characteristics of individual leaves, and information on soil characteristics are treated as the model parameters.

3.3. Model Parameters

To apply the multilayer model, we used the parameters shown in Table 1. The forests at the Kog-Ma experimental watershed in Thailand and the O Thom I experimental watershed in Cambodia differ. Therefore, we used parameters that are relevant to the canopy structure and leaf characteristics of Cambodian lowland dry evergreen forest.

Seasonal variations in the LAI were measured near the tower site (Table 2). The canopy was divided into 50 equal layers. Vertical profiles of the nondimensional leaf area density (a(z)h/LAI) were determined as a D-type beta distribution, defined as when the upper part of the canopy is denser [7], based on the vegetation structure inventory [13]. Meanwhile, the physical and optical characteristics of individual leaves, such as the drag and bulk coefficients for sensible heat, transmissivity and reflectivity of solar radiation, and photosynthetically active radiation were assumed to be the same as those reported by Tanaka et al. [7] for their hill evergreen forest site in Thailand.

The parameters relating to individual leaves or the canopy structure, such as slope (m), maximum rate of carboxylation at 25ºC (V_{cmax25}), and water storage capacity on the upper side of an individual leaf (W_{Umax}), tended to have a large effect on the simulation results in the parameter-tuning analysis of Tanaka et al. [7]. However, the characteristics of these parameters at the Cambodian lowland dry evergreen forest are obscure. Therefore, in this research, these three parameters were fixed throughout the observation period at values at which the annual amount of simulated evapotranspiration became approximately equivalent to the observed result (see Table 1).

Table 2. Seasonal variation in the leaf area index obtained near the observation tower.

	Month												Annual
	1	2	3	4	5	6	7	8	9	10	11	12	
average (3 years)*	3.95	4.01	3.96	3.89	4.06	4.32	4.39	4.26	4.10	3.98	3.87	3.85	4.05
2004	4.18	4.21	4.07	3.91	3.99	4.26	4.49	4.37	4.03	3.87	3.93	3.94	4.10

*Apr 2003- Mar 2006

Because the soil water content at 100 cm depth maintained a wet soil condition during the dry season [28], and the ground water level was approximately 3 m below the ground surface even at the end of the dry season [29], we assumed that drying within the soil did not occur during the analysis period. Evaporation from the forest floor was considered to be negligible because the forest canopy was very dense, and thus radiation transfer to the soil surface was assumed to be very small.

Rainfall at the tower site was measured from 25 April 2004. Generally, the period from January to April corresponds to the dry season [13,30]; because we did not record rainfall for this period, we used input values of zero rainfall for our multilayer model during this period. We do not consider this assumption to contradict because the ground water level gradually decreased during the period in 2004 for which data are lacking [28]. The time interval for our multilayer model simulation was 10 min, corresponding to the interval between individual meteorological measurements.

4. RESULTS AND DISCUSSIONS

4.1. Seasonal Variation in Each Meteorological Component

Figure 2 shows the seasonal variations in observed and estimated components averaged (or integrated) every three days. The seasonal variations in solar elevation at noon calculated from the latitude and longitude of the tower site

(SE), and observed S↓ and S↑, are shown in Figure 2 (a). The solar elevation was maximal in April and August, corresponding to the late dry season and the middle of the rainy season, respectively. The downward shortwave radiation during the rainy season was almost the same as that during the dry season. The integrated values of downward and upward shortwave radiation in 2004 were 6.84 and 0.76 GJ m^{-2} $year^{-1}$, respectively. The mean value of albedo during the observation period was 0.111. Figure 2(b) shows the seasonal variation in L↓ and L↑ estimated from the meteorological elements (see Appendix). Figure 2(c) presents the seasonal variation in air temperature observed at a height of 34 m (TD_1). The lowest air temperature was during the early dry season (<25ºC), and the highest was during the late dry season (>30ºC). In contrast, temperatures did not fluctuate much during the rainy season, with an average value of approximately 27ºC. Figure 2(d) gives the vapor pressure deficit (VPD) at a height of 34 m, which indicates the dryness of the air. The figure shows a clear tendency of high VPD during the dry season (>5 hPa) and low VPD throughout the rainy season (<5 hPa). Figure 2 (e) illustrates the seasonal variation in rainfall (P) from 25 April onward. Rainfall exceeded 200 mm over 3 $days^{-1}$ twice during the rainy season. In addition, a short dry season, namely a period with little rain during the rainy season [31], was clearly observed between the end of June and the middle of July. The dry season commenced in the middle of October with very little subsequent rainfall. Figure 2 (f) shows the seasonal variation in wind velocity observed at a height of 36 m (u). Although we expected that wind velocity would be high during the rainy season, we did not observe any clear seasonal variation in wind velocity.

From these results, we roughly classified three seasons for the study watershed: 1) an early dry season with cool, dry air and little rain; 2) a late dry season with hot, dry air and little rain; and 3) a rainy season with much rain. Each period corresponded to January to the middle of February and the end of October to December, the end of February to the beginning of May, and the middle of May to the middle of October, respectively.

4.2. Observational Results for Evapotranspiration by the EBBR Method

Figure 3 shows the seasonal variation in evapotranspiration calculated by the EBBR method. The vertical bars show the standard deviations in each period. When seasonal variation in evapotranspiration in 2004 was considered, the average value of evapotranspiration during the rainy season (middle of May to middle of October) was 2.3 mm day^{-1}, which was less than the average value during the dry season (3.5 mm day^{-1}). Additionally, peaks in the evapotranspiration rate were also observed at the late dry season (7 March 2004: 5.7 mm day^{-1}) and the beginning of the dry season (29 October 2004: 5.5 mm day^{-1}).

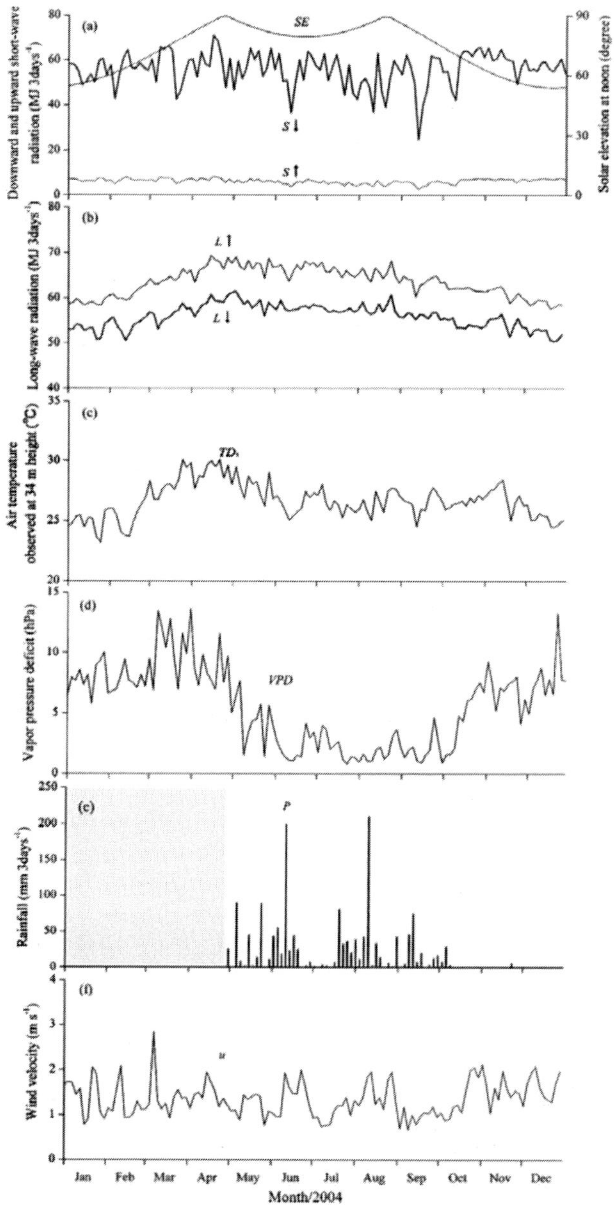

Figure 2. Seasonal variation in (a) solar elevation at noon (SE), downward shortwave radiation (S↓), and upward shortwave radiation (S↑); (b) downward longwave radiation (L↓) and upward longwave radiation (L↑); (c) air temperature (TD_1); (d) vapor pressure deficit (VPD); (e) rainfall (P), shading indicates a period lacking data; and (f) wind velocity (u). All values were averaged or integrated every three days.

As shown in Figures 2(a) and (b), no large seasonal variations in S↓, S↑, L↓, and L↑ between the dry season and rainy season were found at this site. The highest temperature was found in the first half of the dry season, followed by the rainy season, and then the latter half of the dry season; this seasonal order differed from the seasonal variation in observed evapotranspiration. On the other hand, as shown in Figure 2(d), we found a clear difference in VPD between the dry and rainy seasons. These results suggest that larger evapotranspiration in the dry season than in the rainy season originated from the seasonal variation in the VPD.

Stomatal aperture at the leaf level is modeled as the increase in VPD leading to stomatal closure [32]. Based on this concept, increasing VPD seems to lead to decreasing evapotranspiration. Meanwhile, transpiration at the community level is composed of equilibrium evaporation and imposed evaporation, defined by the proportionality relation to effective radiation, and multiplication of VPD and stomatal conductance, respectively [33]. Based on this definition, the increment of VPD may cause an increase in transpiration. Furthermore, even in the dry season, deeper soil remained wet in our experimental watershed [11,28,29]. Therefore, trees could obtain sufficient water even during the dry season with high VPD; thus, little stomatal closure may have occurred at the study site. Therefore, the depression rate of stomatal conductance accompanying the increment in VPD would not be so large, and as a result, the imposed evaporation may increase with the high VPD.

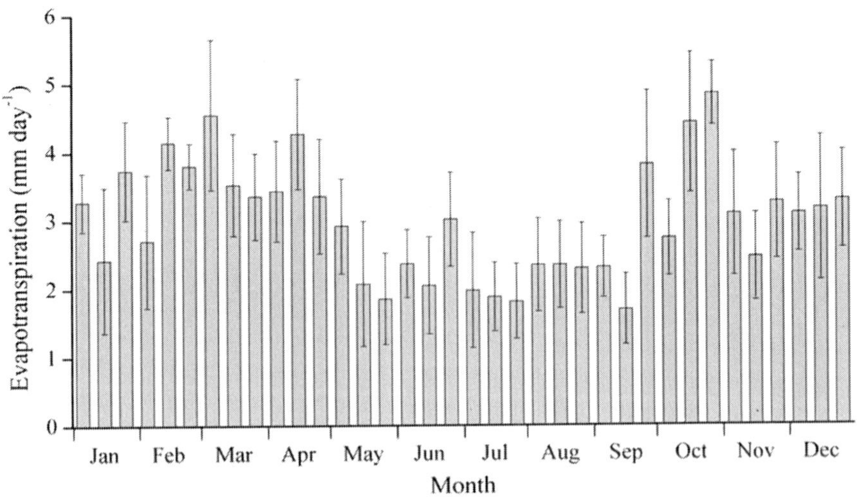

Figure 3. Daily amount of evapotranspiration by the EBBR method in 2004 shown as monthly based 10-day averages. Vertical bars indicate standard deviations for each observation period.

4.3. Comparison Between Observed and Simulated Evapotranspiration

Figure 4 presents the evapotranspiration rates obtained from the observations through EBBR method and from the multilayer model. Each point shows the observed and simulated daily evapotranspiration rates averaged over three days.

In this study, we used the following types of parameters in the multilayer model: 1) those obtained from direct measurement of a lowland dry evergreen forest in Cambodia (e.g., LAI, canopy height); 2) those identified in the hill evergreen forest at Kog-Ma, Thailand (e.g., physical and optical characteristics of an individual leaf); and 3) those determined based on the annual amount of simulated evapotranspiration's becoming approximately the same as the observed result (m, V_{cmax25}, and W_{Umax}). Therefore, the model did not necessarily represent the present condition of the lowland dry evergreen forest in O Thom I. However, the model substantially reproduced the observed result of larger evapotranspiration in the dry season by the EBBR method. Therefore, many of the parameters assumed in this research may have roughly expressed the characteristics of the Cambodian lowland dry evergreen forest.

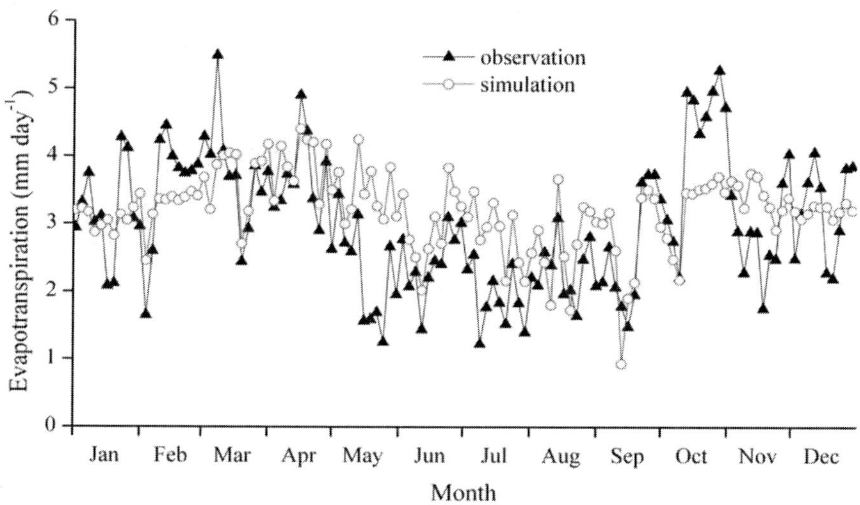

Figure 4. Comparison between observed (closed triangles) and simulated (open circles) daily amounts of evapotranspiration in 2004 averaged every three days.

Because some parameter values were determined by the simulated total amount of evapotranspiration nearing the observation result, the values estimated by the multilayer model were sometimes more or less than observed values. The observed evapotranspiration exceeded the simulated values during the late dry season (middle of February to the beginning of March) and at the beginning of the dry season (beginning of October), as noted in the first paragraph of Section 4.2. These periods coincided mostly with the peak of seasonal variation in evapotranspiration. Furthermore, during the rainy season, the model results overall tended to exceed the observed evapotranspiration.

Seasonal variation in the physiological characteristics of individual leaves may be one explanation for why the model did not reproduce the two peaks of observed evapotranspiration during the dry season and overestimated evapotranspiration in the rainy season. In the model simulation conducted in this study, parameters such as the photosynthetic rate and stomatal aperture of each leaf were assumed to be fixed values. However, the photosynthetic rate is known to reach maximum values at the time of foliation and then decline with advancing leaf age [34–36]. Such a phenological feature has also been recognized in photosynthesis observations of evergreens in other tropical regions [37,38]. The stomatal conductance of evergreen trees has also been demonstrated to fluctuate seasonally due to variation in environmental factors such as meteorological conditions and soil aridity in the seasonal tropics [39–43]. Therefore, the parameters concerning the assimilation rate and stomatal aperture at our study site might also change with the maturity and senescence of leaves.

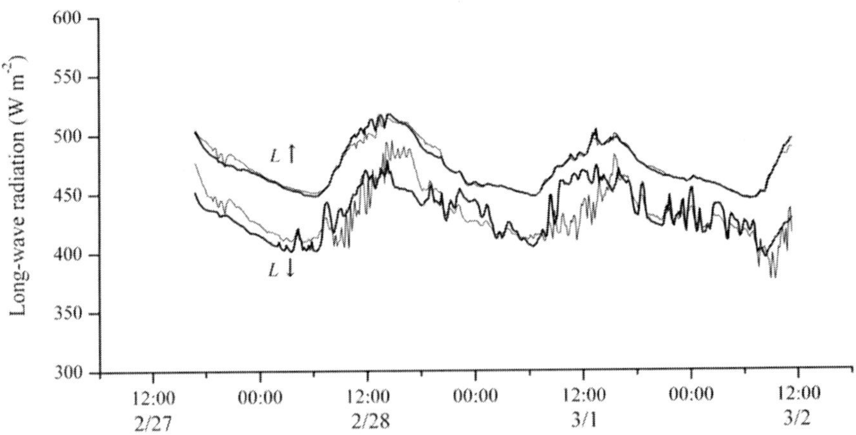

Figure 5. Time series of longwave radiation from 27 February to 2 March 2006. L↓ and L↑ indicate downward and upward longwave radiation, respectively, and the thick and thin lines indicate observed and estimated values, respectively.

Leaf dynamics clearly display seasonal changes in the evergreen forests of the study area. Investigations of litter fall and the LAI showed that leaf abscission and foliation became active at the beginning and late parts of the dry season

(E. Ito, unpublished data). Therefore these phenological characteristics thus probably did not agree with fixed assumptions for parameters of photosynthesis and evapotranspiration throughout the year. However, many tree species with various phenologies intermingle with the evergreen forest neighboring the meteorological observation site; on the whole, the seasonal variation in physiological characteristics for the various species has not yet been clarified quantitatively. Therefore, we presently do not have enough information to explain the peaks of evapotranspiration during the dry season by the physiology of individual leaves alone or to reflect this issue in the model simulation. However, on the basis of the above discussion, we suggest that seasonal variation in the physiological characteristics of individual leaves of various tree types in the surrounding forest had a large effect on the evapotranspiration in the O Thom I experimental watershed. The observational findings of the transpiration peak during the dry season and the higher daily amount of evapotranspiration in the dry season than in the rainy season would be strongly affected by those seasonal variations.

A next research step will be to further investigate the characteristics of tree physiology postulated on the basis of the model simulation in this study.

5. CONCLUSIONS

This study confirmed the following findings. From meteorological observations, we could roughly classify three seasons for a lowland dry evergreen forest in central Cambodia: an early dry season, a late dry season, and a rainy season. We then used the energy balance Bowen ratio (EBBR) method to observe the evapotranspiration rate. The daily evapotranspiration amount was higher in the dry season (3.5 mm day^{-1}) than in the rainy season (2.3 mm day^{-1}). The high vapor pressure deficit during the dry season likely had a large effect on the higher evapotranspiration during the dry season. The finding of higher evapotranspiration in the dry season than in the rainy season was also reproduced by the multilayer model of Tanaka et al. [7].

However, our results also showed differences between the simulation and observation. During the beginning and late parts of the dry season, observed evapotranspiration was greater than simulated evapotranspiration. In contrast to the dry season, simulated evapotranspiration during the rainy season tended to be larger than observed evapotranspiration. The above results suggest that one cause of the quantitative difference in observed and simulated evapotranspiration in some periods was the seasonal variation in physiological characteristics such as photosynthesis and/or transpiration, which we treated as constant parameters in the multilayer model simulation.

Further investigation of physiological and phenological features accompanied by continuous observation of evapotranspiration in lowland dry evergreen forest of Cambodia would help clarify the above suggestions.

6. ACKNOWLEDGMENTS

This study was partly funded by the Ministry of the Environment, Japan (Global Environment Research Coordination System). We also thank Mr. Ty Sokhun, Director General of the Forestry Administration of Cambodia, and our Cambodian colleagues of the Forest and Wildlife Science Research Institute, Cambodia.

7. APPENDIXES

7.1. Estimation of Downward Longwave Radiation

Downward longwave radiation was a necessary input for the multilayer model. However, only short-term observations of upward longwave radiation (L↑) and downward longwave radiation (L↓) were carried out in this study. For this reason, it was necessary to presume downward longwave radiation (L↓) throughout the period. We explain the method that we used to presume L↓ in this section.

Net radiation (Rn) is composed of shortwave and longwave radiation, as given in Equation (4):

$$Rn = S\downarrow + S\uparrow + L\downarrow + L\uparrow \quad (4)$$

Here, with radiation energy from the black body proportionate to the fourth power of the absolute temperature, the following equation is approved at the canopy surface:

$$L\uparrow = \sigma \times K^4, \quad (5)$$

where σ is the proportionality constant value and K is the absolute temperature of the canopy surface. The temperature of the canopy surface was not measured; therefore the temperature observed at 34 m height was substituted into Equation (5) instead of the canopy surface temperature. Equation (6) was then rearranged as follows:

$$L\uparrow = \sigma \times (TD_1 + 273.15)^4, \quad (6)$$

where TD_1 is the temperature observed at a height of 34 m. Then the following Equation (7) was obtained with reference to Equations (4) and (6):

$$L\downarrow = Rn - S\uparrow - S\downarrow - \sigma(TD_1 + 273.15)^4. \quad (7)$$

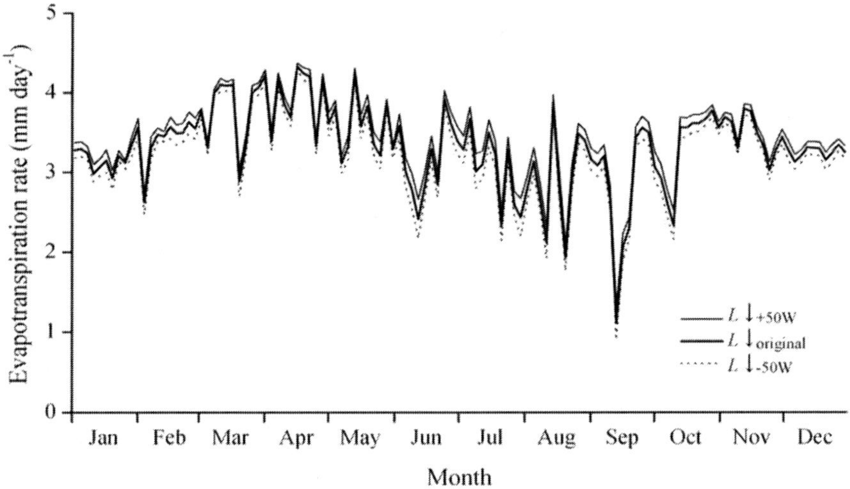

Figure 6. The effect of downward longwave radiation (L↓) on evapotranspiration during the daytime. The legends show the calculated evapotranspiration rate using L↓ ($L\downarrow_{original}$) estimated from Equations (7), with 50 W m^{-2} added to L↓ ($L\downarrow_{+50W}$), and with 50 W m^{-2} subtracted from L↓ ($L\downarrow_{-50W}$), respectively.

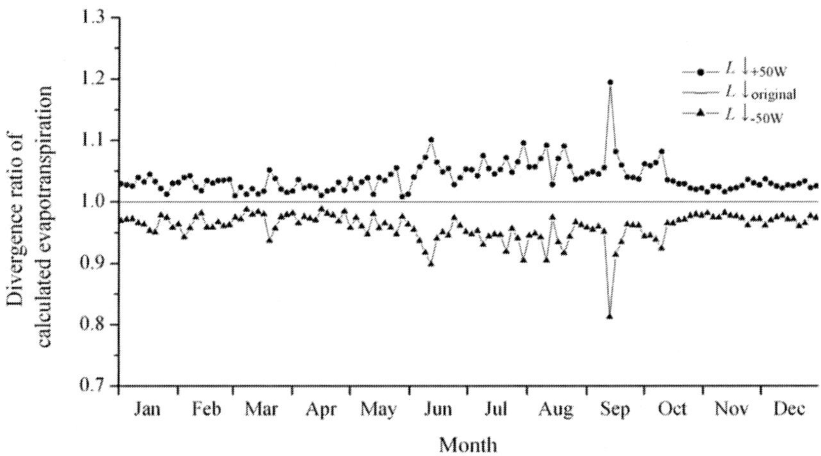

Figure 7. The divergence ratio, calculated by the divergence of the evapotranspiration with $L\downarrow_{+50W}$ and $L\downarrow_{-50W}$ from the evapotranspiration with the original L↓. The legends are the same as in Figure 6.

The estimated L↑ and L↓ obtained from Equations (6) and (7) were verified by observed L↑ and L↓ obtained during the intensive observation period from 27 February to 2 March 2006 (Figure 5). When σ was defined as 5.61×10^{-8} (W m^{-2}

K^{-4}), the coefficient of determination between observed and estimated L↓ was 0.54. Maximum estimation error in L↓ was about ±50 W m^{-2}, and the difference between integral values of observed and estimated L↓ during the daytime was about 3%.

7.2. Effect of L↓ Error on Simulated Evapotranspiration

As mentioned above, the maximum estimation error in L↓ was approximately ±50 W m^{-2}. To verify that the effect of the L↓ estimation error on the calculated evapotranspiration rate changed between −50 and +50 W m^{-2}, numerical simulation was performed. Figure 6 shows the seasonal variations in evapotranspiration obtained from the simulation results by using (1) L↓ estimated from $\sigma = 5.61 \times 10^{-8}$ ($L{\downarrow}_{original}$), (2) an increase of 50 W m^{-2} from $L{\downarrow}_{original}$ ($L{\downarrow}_{+50W}$), and (3) a decrease of 50 W m^{-2} from $L{\downarrow}_{original}$ ($L{\downarrow}_{-50W}$). The three simulations resulted in the same tendency of seasonal patterns of evapotranspiration.

Figure 7 shows the ratio of divergence in evapotranspiration obtained by the cases of $L{\downarrow}_{+50W}$ and $L{\downarrow}_{-50W}$ from the basic simulation result ($L{\downarrow}_{original}$). During the period when the net radiation was small, the estimation error of L↓ had a relatively large effect on the simulated radiation output, but during the other periods, the estimation error contributed less than 5%. However, when the accuracy of daily-integrated L↓ was considered, very little difference was observed between the numerical simulations illustrated in Figure 7. From the above, we concluded that the estimation method of L↓ used in this study was appropriate.

REFERENCES

1. A. Nalampoon, "National forest policy review, Thailand," in P. B. Durst Ed., "An overview of forest policies in Asia," FAO, Bangkok, pp. 293–311, 2003.
2. E. Savet and T. Sokhun, "National forest policy review, Cambodia," in P. B. Durst Ed., "An overview of forest policies in Asia," FAO, Bangkok, pp. 93–108, 2003.
3. B. Krishnapillay, "Silviculture and management of teak plantations," Unasylva, Vol. 201, pp. 14–21, 2000.
4. A. Mittelman, "Teak planting by smallholders in Nakhon Sawan, Thailand," Unasylva, Vol. 201, pp. 62–65, 2000.
5. R. T. Pinker, O. E. Thompson, and T. F. Eck, "The energy balance of a tropical evergreen forest," Journal of Applied Meteorology, Vol. 19, pp. 1341–1350, 1980.

6. P. Attarod, M. Aoki, D. Komori, T. Ishida, K. Fukumura, S. Boonyawat, P. Tongdeenok, M. Yokoya, S. Punkngum, and T. Pakoktom, "Estimation of crop coefficients and evapotranspiration by meteorological parameters in a rain-fed paddy rice field, cassava and teak plantations in Thailand," Journal of Agricultural Meteorology, Vol. 62, pp. 93–102, 2006.
7. K. Tanaka, H. Takizawa, N. Tanaka, I. Kosaka, N. Yoshifuji, C. Tantasirin, S. Piman, M. Suzuki, and N. Tangtham, "Transpiration peak over a hill evergreen forest in northern Thailand in the late dry season: Assessing the seasonal changes in evapotranspiration using a multilayer model," Journal of Geophysical Research, Vol. 108 (D17), pp. 4533, 2003.
8. T. Kume, H. Takizawa, N. Yoshifuji, K. Tanaka, C. Tantasirin, N. Tanaka, and M. Suzuki, "Impact of soil drought on sap flow and water status of evergreen trees in a tropical monsoon forest in northern Thailand," Forest Ecology and Management, Vol. 238, pp. 220–230, 2007.
9. K. Tanaka, H. Takizawa, T. Kume, J. Xu, C. Tantasirin, and M. Suzuki, "Impact of rooting depth and soil hydraulic properties on the transpiration peak of an evergreen forest in northern Thailand in the late dry season," Journal of Geophysical Research, Vol. 109, D23107, 2004.
10. N. Tanaka, T. Kume, N. Yoshifuji, K. Tanaka, H. Takizawa, K. Shiraki, C. Tantasirin, N. Tangtham, and M. Suzuki, "A review of evapotranspiration estimates from tropical forests in Thailand and adjacent regions," Agricultural and Forest Meteorology, Vol. 148, pp. 807–819, 2008.
11. Y. Ohnuki, C. Kimhean, Y. Shinomiya, and J. Toriyama, "Distribution and characteristics of soil thickness and effects upon water storage in forested areas of Cambodia," Hydrological Processes, Vol. 22, pp. 1272–1280, 2008.
12. N. Kabeya, A. Shimizu, T, Nobuhiro, and K. Tamai, "Preliminary study of flow regimes and stream water residence times in multi-scale forested watersheds of central Cambodia," Paddy and Water Environment, Vol. 6, pp. 25–35, 2008.
13. A. Shimizu, N. Kabeya, T. Nobuhiro, T. Kubota, Y. Tsuboyama, E. Ito, M. Sano, S. Chann, and N. Keth, "Runoff characteristics and observations on evapotranspiration in forest watersheds, central Cambodia," in H. Sawada et al. Ed., "Forest Environments in the Mekong River Basins," Springer, Tokyo, pp. 135–146, 2007.

14. J. Toriyama, S. Ohta, M. Araki, M. Kanzaki, S. Khorn, P. Pith, S. Lim, and S. Pol, "Soils under different forest types in the dry evergreen forest zone of Cambodia: Morphology, physicochemical properties, and classification," in H. Sawada et al. Ed., Forest Environments in the Mekong River Basins," Springer, Tokyo, pp. 241–253, 2007.
15. Y. Ohnuki, C. Kimhean, Y. Shinomiya, S. Sor, J. Toriyama, and S. Ohta, "Apparent change in soil depth and soil hardness in forest areas in Kampong Thom province, Cambodia," in H. Sawada et al. Ed., "Forest Environments in the Mekong River Basins," Springer, Tokyo, pp. 263–272, 2007.
16. A. Shimizu, T. Shimizu, Y. Miyabuchi, and Y. Ogawa, "Evapotranspiration and runoff in a forested watershed, western Japan," Hydrological Processes, Vol. 17, pp. 3125–3139, 2003.
17. S. Hattori, "Explanation on derivation process of equations to estimate evapotranspiration and problems on the application to forest stand," Bulletin of the Forestry and Forest Products Research Institute, Japan, Vol. 332, pp. 139–165, 1985.
18. J. C. Kaimal and J. J. Finnigan, "Atmospheric boundary layer flows: Their structure and measurement," Oxford University Press, New York, 1994.
19. W. Brutsaert, Ed., "Evaporation into the atmosphere: theory, history, and applications," D. Reidel Publishing Company, 1982.
20. K. Tamai, T. Nobuhiro, A. Shimizu, N. Kabeya, S. Chann, and K. Nang, "Characteristics of atmospheric stability above an evergreen forest in central Cambodia," Hydrological Processes, Vol. 22, pp. 1267–1271, 2008.
21. G. D. Farquhar, S. V. Caemmerer, and J. A. Berry, "A biochemical model of photosynthetic CO_2 assimilation in leaves of C_3 species," Planta, Vol. 149, pp. 78–90, 1980.
22. J. T. Ball, "An analysis of stomatal conductance," Ph.D thesis, Stanford University, Stanford, California, 1988.
23. T. Watanabe, "The bulk transfer coefficients over a vegetated surface based on K-theory and 2nd-order closure model," Journal of the Meteorological Society of Japan, Vol. 71, pp. 33–42, 1993.
24. N. R. Wilson and E. H. Shaw, "A higher order closure model for canopy flow," Journal of Applied Meteorology, Vol. 16, pp. 1197–1205, 1977.
25. K. Tanaka, "Multi-layer model of CO_2 exchange in a plant community coupled with the water budged of leaf surfaces," Ecological Modelling, Vol. 147, pp. 85–104, 2002.

26. K. Tanaka, Y. Kosugi, N. Ohte, S. Kobashi, and A. Nakamura, "Model of CO_2 flux between a plant community and the atmosphere, and simulation of CO_2 flux over a planted forest," (in Japanese with an English abstract) Japanese Journal of Ecology, Vol. 48, pp. 265–286, 1998.
27. J. Ross, "Radiative transfer in plant communities," in J. L. Monteith Ed., "Vegetation and the atmosphere," Vol. 1, Academic Press, California, pp. 13–55, 1975.
28. M. Araki, J. Toriyama, S. Ohta, M. Kanzaki, E. Ito, B. Tith, S. Pol, S. Lim, S. Khorn, P. Pith, and S. Det, "Soil moisture conditions in four types of forests in Kampong Thom, Cambodia," in H. Sawada et al. Ed., "Forest environments in the Mekong river basins," Springer, Tokyo, pp. 254–262, 2007.
29. M. Araki, A. Shimizu, J. Toriyama, E. Ito, N. Kabeya, T. Nobuhiro, B. Tith, S. Pol, S. Lim, S. Khorn, P. Pith, S. Det, S. Ohta, and M. Kanzaki, "Changes of vertical soil moisture conditions of a dry evergreen forest in Kampong Thom, Cambodia," in H. Sawada et al. Ed., "Forest environments in the Mekong river basins," Springer, Tokyo, pp. 112–124, 2007.
30. N. Kabeya, A. Shimizu, S. Chann, Y. Tsuboyama, T. Nobuhiro, N. Keth, and K. Tamai, "Stable isotope studies of rainfall and stream water in forest watersheds in Kampong Thom, Cambodia," in H. Sawada et al. Ed., "Forest environments in the Mekong river basins," Springer, Tokyo, pp. 125–134, 2007.
31. J. Matsumoto, "Seasonal transition of summer rainy season over Indochina and adjacent monsoon region," Advances in Atmospheric Sciences, Vol. 14, pp. 231–245, 1997.
32. J. B. Stewart, "Modelling surface conductance of pine forest," Agricultural and Forest Meteorology, Vol. 43, pp. 19–35, 1988.
33. K. G. McNaughton and P. G. Jarvis, "Predicting effects of vegetation changes on transpiration and evaporation," in T. T. Kozlowski Ed., "Water deficits and plant growth," Academic Press, New York, Vol. 7, pp. 1–47, 1983.
34. D. S. Ellsworth, "Seasonal CO_2 assimilation and stomatal limitations in a Pinus taeda canopy," Tree Physiology, Vol. 20, pp. 435–445, 2000.
35. S. Mediavilla and A. Escudero, "Photosynthetic capacity, integrated over the lifetime of a leaf, is predicted to be independent of leaf longevity in some tree species," New Phytologist, Vol. 159, pp. 203–211, 2003.

36. Q. Han, T. Kawasaki, T. Nakano, and Y. Chiba, "Spatial and seasonal variability of temperature responses of biochemical photosynthesis parameters and leaf nitrogen content within a Pinus densiflora crown," Tree Physiology, Vol. 24, pp. 737–744, 2004.
37. K. Kitajima, S. S. Mulkey, and S. J. Wright, "Seasonal leaf phenotypes in the canopy of a tropical dry forest: Photosynthetic characteristics and associated traits," Oecologia, Vol. 109, pp. 490–498, 1997.
38. K. Kitajima, S. S. Mulkey, and S. J. Wright, "Decline of photosynthetic capacity with leaf age in relation to leaf longevities for five tropical canopy tree species," American Journal of Botany, Vol. 84, pp. 702–708, 1997.
39. F. C. Meinzer, G. Golostein, N. M. Holbrook, P. Jackson, and J. Cavelier, "Stomatal and environmental control of transpiration in a lowland tropical forest tree," Plant, Cell and Environment, Vol. 16, pp. 429–436, 1993.
40. P. B. Reich, "Phenology of tropical forests: Patterns, causes, and consequences," Canadian Journal of Botany, Vol. 73, pp. 164–174, 1994.
41. J. I. Pitman, "Ecophysiology of tropical dry evergreen forest, Thailand: measured and modelled stomatal conductance of Hopea ferrea, a dominant canopy emergent," Journal of Applied Ecology, Vol. 33, pp. 1366–1378, 1996.
42. D. S. Thomas, D. Eamus, and S. Shanahan, "Influence of season, drought and xylem ABA on stomatal responses to leaf-to-air vapour pressure difference of trees of the Australian wet-dry tropics," Australian Journal of Botany, Vol. 48, pp. 143–151, 2000.
43. K. Kitajima, S. S. Mulkey, M. Samaniego, and S. J. Wright, "Decline of photosynthetic capacity with leaf age and position in two tropical pioneer tree species," American Journal of Botany, Vol. 89, pp. 1925–1932, 2002.

Chapter 7

Determination of Evapotranspiration and Water Use Efficiency in Crop Production

Kazeem O. Rauff[1*], Saheed A. Shittu[2]

[1]Federal University, Kashere, Nigeria

[2]Federal College of Animal Health & Production Technology, Ibadan, Nigeria

ABSTRACT

The main concern in cultivating crops has always been water availability. To increase food production, water plays a major role after securing a large portion of land area. Knowledge of the factors influencing crop water use efficiency and hope to increase the efficiency has continued to be an objective in many modern studies. The gap between irrigation supply and demand is increasing from year to year as a result of an increase in population growing rate and people moving from place to place. Searching for new water may be a too difficult and very expensive process, so the shortest and easiest way is to maximize the water use efficiency throughout optimizing water use efficiency and the first step on that is determination of the actual crop water requirements.

Keywords: Evapotranspiration, Micrometeorological, Bowen Ratio, Microclimate, Turbulent Transport

1. INTRODUCTION

The crop evapotranspiration is defined like the water transferred to atmosphere by plant transpiration and surface evaporation. The evapotranspiration term was proposed by [1] to conceptualize the process of plant transpiration and surface evaporation which occurs simultaneously and no easy way to distinguish them. The evapotranspiration process occurs naturally only if there is inflow of energy in the system, from the sun, atmosphere, or both, and is controlled by the rate of energy in the form of water vapor that spreads from the surface of the Earth [2] . This transfer takes place physically in the forms of molecular and turbulent diffusion.

According to [3] , evaporation is the process whereby liquid water is converted to water vapor (vaporization) and removed from the evaporating surface (vapor removal). Water evaporates from a variety of surfaces, such as lakes, rivers, pavements, soils and wet vegetation. Transpiration, in turn, is the transfer of water from plants through their aerial parts. Water transfer from plant to atmosphere occurs mainly through the stomata through which they pass more than 90% water transpired.

The evapotranspiration can be derived from a range of measurement systems including lysimeters, eddy covariance, Bowen ratio, water balance (gravimetric, neutron meter, other soil water sensing), sap flow, scintillometer and even satellite-based remote sensing and direct modeling [4]. The micrometeorological method of Bowen Ratio Energy Balance (BREB) has been widely applied due to its relative simplicity, practicality, robustness and accuracy. Recently, the Eddy Covariance method has been increasingly applied mainly after the sensors have reduced costs.

The science of micrometeorology, which deals with atmospheric processes near the earth's surface, is providing increased knowledge of the mechanisms controlling our environment. Micrometeorological methods have only recently been applied to agronomic problems. Such application has increased our understanding of the response of crop plants to their physical environment, thereby permitting us to develop improved agronomic practices aimed at achieving more efficient use of water with increased and more economical production of crops [5].

The main aim of this discussion is to summarize some of the key factors in the application of the meteorological methods to agronomic problems. The application of the methods to the problems of increasing water-use efficiency and crop production will be emphasized rather than the details of the methods. This will include a review of some of the basic physical factors controlling the microclimate at any point on the earth's surface and the application of the meteorological approach in studying this microclimate [5].

The meteorological method has a twofold application to agronomic problems. First, it serves as a useful research tool, permitting the study of basic physical processes within the microclimate. Second, it has the potential for direct application to farm management practices, such as measurement of the evaporative loss of water and determination of when to irrigate. The expenditure of considerable time, effort and money in the development of these methods is fully justified because only by gaining an increased understanding of the natural physical environment of soils and plants and their response to this environment can we hope to understand and predict plant responses to microclimatic conditions. The resulting information will permit us to develop methods of altering the natural environment to increase crop production and water-use efficiency [5].

2. BOWEN RATIO ENERGY BALANCE (BREB)

The Bowen-ratio method is one of the most common methods used to determine the fluxes of sensible and latent heat. The method is based on Bowen-ratio similarity and the energy balance equation. The energy balance equation provides a sound theoretical basis for estimating evaporation from natural surfaces. For a natural surface the complete equation can be written as

$$Q_s^* = Q_H + Q_E + Q_G + Q_V + Q_W + Q_P \tag{1}$$

where:

Q_H = turbulent transfer of sensible heat flux which is responsible for heating the atmosphere from the surface up to some 100 m during the day, except for days with strong convection,

Q_E = latent heat energy loss by evapotranspiration,

Q_G = vertical heat flux downward into the ground,

Q_V = horizontal heat flux, or advection, out of the region,

Q_W =net heat loss due to vertical water storage exchange,

Q_P = net energy used in such natural processes as photosynthesis.

The vertical water storage exchange term refers to the transfer through evaporation and precipitation at temperature different than the mean surface temperature. The relative magnitudes of terms in equation above are highly dependent on the space and time scales under consideration, the characteristics and homogeneity of the surface, and atmospheric conditions.

The crop evapotranspiration is estimated by use of the BREB technique from latent heat flux density which is derived from energy balance equation (Equation (2)). For many practical purposes the last three terms are negligible and the equation reduces to [6] :

$$-Q_s^* = Q_H + Q_E + Q_G \tag{2}$$

where: $-Q_s^*$ is the net radiation, Q_H is the sensible heat flux which is responsible for heating the atmosphere from the surface up to some 100 m during the day, except for days with strong convection, Q_E is the latent heat energy loss by evapotranspiration and Q_G is the vertical heat flux downward into the ground.

On the other hand [4] lists some potential advantages of BREB method, which are:

- non-destructive, direct sampling of the turbulent boundary layer;
- no aerodynamic data are required;
- simple measurement of temperature and vapor pressure at two heights;
- can measure ET over both potential and non-potential surfaces;
- gradient-based fluxes are averaged over a medium sized area (200 - 100,000 m^2); automated.

Additionally it is relative simplicity of method made it widely applied [7] -[11] .

The Bowen ratio method can be used to access the apportionment of the remaining available energy between the sensible heat term Q_H and the latent heat term Q_E. The Bowen ratio can be determined from the gradients of temperature and vapor pressure above the evaporating surface, using either the equation quoted by [6]:

$$B = \frac{Q_H}{Q_E} = \frac{0.62P}{1000}\frac{T_2 - T_1}{e_2 - e_1} \tag{3}$$

or that by [12] :

$$B = \left[\left(\frac{S}{\gamma}+1\right)\left(\frac{T_{w2}-T_{w1}}{T_2-T_1}\right)-1\right]^{-1} = \left[\left(\frac{S}{\gamma}+1\right)\left(\frac{\Delta T_w}{\Delta T}\right)-1\right]^{-1} \tag{4}$$

where:

B = Bowen ratio,

P = atmospheric pressure (mb),

T_2, T_1 = potential temperature (˚C) at heights z_1 and z_2,

T_{w2}, T_{w1} = wet bulb temperatures at z_1 and z_2,

S = $\frac{\partial e}{\partial T_W}\left[\text{mb}/(^\circ\text{C})^{-1}\right]$ at the mean wet bulb temperature $\overline{T}_W$,

γ = psychrometric constant,

$$B = \frac{c_p}{\lambda}\frac{\Delta T}{\Delta q} = \frac{c_p}{\lambda}\frac{p}{0.622}\frac{\Delta T}{\Delta e} = \gamma\frac{\Delta T}{\Delta e} \quad (5)$$

where the psychrometric constant is $\gamma = 0.667\ \text{K}\cdot\text{hPa}^{-1}$, the pressure p = 1000 hPa, the specific heat for constant pressure $c_p = 1004.834\left[\text{J}\cdot\text{K}^{-1}\cdot\text{kg}^{-1}\right]$, the evaporation heat of water $\lambda = 2500827 - 2360(T - 273.1)\left[\text{J}\cdot\text{kg}^{-1}\right]$,

the air density $\rho = \frac{p[\text{hPa}]\cdot 100}{R_L \cdot T_v}\left[\text{kg}\cdot\text{m}^{-3}\right]$, and $t = 20^\circ\text{C}$,

From both Equations (1) and (2), the sensible and latent heat fluxes can be determined:

$$\frac{-Q_s^*}{Q_E} = \frac{Q_H}{Q_E} + \frac{Q_E}{Q_E} + \frac{Q_G}{Q_E} \quad (6)$$

$$Q_E = \frac{\left(-Q_s^* - Q_G\right)}{B+1} \quad (7)$$

$$\frac{-Q_s^*}{Q_E} = B + 1 + \frac{Q_G}{Q_E} \quad (8)$$

$$\frac{-Q_s^*}{Q_E} - \frac{Q_G}{Q_E} = B + 1 \quad (9)$$

$$\frac{1}{Q_E}\left(-Q_s^* - Q_G\right) = B + 1 \quad (10)$$

$$Q_H = \left(-Q_s^* - Q_G\right)\frac{B}{B+1} \quad (11)$$

The latent heat flux can also be expressed as $Q_E = \lambda E$, where $\lambda \approx 2.45\times10^6\ \mathrm{J\cdot kg^{-1}}$ is the latent heat of evaporation/condensation and E is the rate of evaporation/condensation. The accuracy of the Bowen ratio has been discussed by [13] (Tanner, 1960) among others. It is apparent that Equations (7) and (11) do not include the wind velocity and do not prescribe a certain difference between the measurement heights.

The crucial disadvantage of the Bowen-ratio method is that because of the apparent unclosed energy balance the residual is either added to the net radiation or distributed according to the Bowen ratio to the sensible and latent heat flux. In general, the fluxes determined with the Bowen-ratio method are larger than those determined with the eddy-covariance method.

The sensible heat flux can be directly determined with a sonic anemometer and a correction of the buoyancy flux [14] (Liu and Foken, 2001). The Bowen ratio can be calculated with Equation (3), and the latent heat flux can be calculated with:

$$Q_E = \frac{Q_H}{B} \tag{12}$$

3. METEOROLOGICAL METHODS AND THE PLANT MICROCLIMATE

3.1. The Plant Microclimate

The physical processes of the atmosphere, called meteorological processes, establish the existing climate or microclimate. Although it does not act alone, climate, in turn, determines the regime of soils and plants at a given location. Wind, precipitation, sunshine, temperature, humidity, and soil moisture are the primary factors involved. The profitable production of crops and efficient use of water require a microclimate suitable for plant growth. What makes the microclimate suitable? The meteorological methods are helping us answer this question. Recent studies of the nature of the microclimate have yielded a wealth of information on the interaction of the physical processes of the environment and the response of plants to these processes. A brief review of the results from these studies will be presented here [15] -[18] . The meteorological processes are dynamic, each reacting to the changes in the others. The dynamic balance involves a continual exchange of mass and energy. Of particular importance to us are the exchanges of molecular heat energy, radiant energy, momentum, water vapor, carbon dioxide, and oxygen between the plant or soil surface and the atmosphere. We refer to this exchange as the vertical flux of the respective quantities. The meteorological method currently offers one of the best approaches to the study of these fluxes. It also permits an evaluation of the rates of the associated plant physiological processes such as photosynthesis and transpiration under natural field conditions [5] .

The energy exchange processes are basic to the rest of the dynamic system. The sun provides the source of radiant energy called solar radiation. Of the total solar energy reaching the outer edge of the earth's atmosphere, about, 65% is available at the earth's surface for processes requiring an energy input. The other 35% is reflected by the atmosphere and the earth's surface. The solar energy absorbed by the earth causes it to be warmed. The heated surface emits thermal radiation back into space, resulting in a loss of energy and a cooling of the surface. We refer to the portion of the radiant energy absorbed at the surface and remaining after the emitted thermal radiation portion is subtracted as the net radiation component. This is the important component so far as the energy-utilizing processes at the earth's surface are concerned because it is partitioned among the various physical and physiological processes existing here [5] .

Net radiation can be measured with reliable instruments. In the absence of direct measurements, net radiation can be obtained using the radiation balance equation if other components are known. The radiation in the atmosphere is divided into shortwave (solar) radiation and long-wave (heat) radiation [19] . Radiation balance at the surface of the Earth is composed of four spectral radiant fluxes; shortwave down welling irradiation from the Sun $(K\downarrow)$in the wavelength range 0.15 - 4 mm, the amount of energy that is reflected$(K\uparrow)$, the incoming long-wave in the wavelength range from (4 - 100 mm) radiation from the atmosphere$(I\downarrow)$; and the amount of long-wave radiation emitted from the surfaces $(I\uparrow)$[19] . The net radiation at the ground surface is given by:

$$Q_S^* = K\downarrow + K\uparrow + I\uparrow + I\downarrow \tag{13}$$

From Equation (13) above we see that the net radiation is the sum of the shortwave down-welling radiation mainly from the sun (global radiation), the long-wave down welling infrared (heat) radiation emitted by clouds, aerosols, and gases, the shortwave up-welling reflected (solar) radiation, and the long wave up-welling infrared (heat) radiation. The shortwave radiation can be divided into the diffuse radiation from the sky and direct solar radiation. The down-welling radiation may be greater if clouds appear in a previously clear sky and the ground surface has cooled. It is only in the case of fog, that up-welling and down-welling radiations are equal. The up-welling long wave radiation is greater than the down-welling long wave radiation, because of the earth's surface is warmer than clouds and aerosols. For clear sky without clouds and dry air, the radiation temperature is approximately? 55°C [19] .

Expressing this in terms of an energy balance, we can say that the net radiation is equal to the total incoming solar radiation minus that which is reflected and that which is lost by radiation from the surface. We can further say that the net radiation is partitioned among the components: 1) sensible heat, utilized in the heating of the air; 2) latent heat, utilized in the evaporation of water; 3) stored heat in the soil and plant materials; and 4) the photochemical energy, utilized in photosynthesis.

As an example of information gained from studies of the microclimate on the interaction of the radiation energy exchanges and the other processes existing within the microclimate, let us consider what happens within a growing crop on a clear summer day. Beginning at sunrise, the soil and plant surfaces are rapidly heated by the absorption of radiant solar energy. Because the air is nearly transparent to shortwave radiation, it does not absorb as much of the radiant energy as do the surfaces. Therefore, the surfaces are much warmer than the air and there is a resulting flow of heat from the surfaces to the air, referred to as sensible heat flux. Because of this, a strong temperature gradient develops very near the surface and the rising of the warmer air parcels causes the air to become unstable. This greatly increases the mixing motion or turbulence of the airstream, which serves to cool the surface and warm the air at even faster rates.

In addition, the solar energy absorbed provides energy for the evaporation of water from the surface. The increased turbulence of the unstable air increases the rate of transfer of water vapor away from the surface, so that there is a rapid drying of soil surfaces and wilting of plants if moisture is limiting.

During photosynthesis, photochemical energy is used by the crop leaves to synthesize organic matter using carbon dioxide from the air. This results in a net transfer of carbon dioxide from the air to the leaves of the plant. Turbulent air motion enhances this transfer rate [5] .

The net result is that the surface gains energy from the sun then loses energy in heating the air, in the evaporation of water and in the flow of heat to the deeper soil layers. This is accompanied by a flow of water vapor away from and a flow of carbon dioxide to, the surface [5].

After sunset, the situation quickly reverses. The surface radiates heat away rapidly and cools to a temperature below that of the air. There is then a flow of heat from the deeper soil layers and from the air to the cooling surface. A strong temperature gradient develops between the air and the soil or plant surface opposite in direction to that existing during the day-time. This condition, called temperature inversion, results in the air mass becoming more stable, with decreased turbulent mixing. Evaporation rapidly decreases, and photosynthesis ceases so there is then an uptake of oxygen and a release of carbon dioxide associated with respiration [5].

The plant and soil surfaces thus act as sources and sinks for the various energy and molecular entities throughout the day. There is a flux of energy and molecular quantities either away from or toward the surface at almost all times, the direction depending upon the daytime conditions. The microclimatic conditions determine the flux rates, but the plant can modify these by stomata closure, shading the ground, and increasing the surface roughness.

Our desire is to be able to quantitatively measure these flux rates on a continuous basis. The meteorological approach serves as a useful tool in enabling us to do this. A complete study of the microclimate involves the measurement of net radiation, total incoming incident radiation, wind speed, air temperature, air humidity, soil temperature, the flow of heat into the soil, the soil moisture content, and the various properties of the surface or plant community itself [5] .

3.2. The Meteorological Approach

The meteorological approach to determining the vertical flux of energy and mass from a crop surface requires the measurement of the respective gradients, with height, of the energy and molecular entities and the determination of a transfer coefficient. The plant and soil surfaces acting as sources and sinks for the various entities establish the concentration or energy differences, which we call gradients, along which the respective entities are transferred.

The transfer rate, or flux, of the particular entity is related to its gradient by a transfer coefficient. This coefficient, which is similar to a diffusion coefficient, characterizes more than any other single parameter the effectiveness of the exchange process [5].

Of particular interest to the micrometeorologist is the flux of carbon dioxide and water vapor associated with photosynthesis and evapotranspiration. The evaporation rate can be calculated under natural field conditions without altering the crop or soil system from measurements of the water vapor gradient and the transfer coefficient. Short period results so necessary in the evaluation of the important aspects of the highly transient system are possible. The meteorological approach also lends itself to mobility. It does, however, require careful application, considerable instrumentation, and involved analysis procedures. Recent advances in instrumentation, electronic technology, and computer developments have enabled use of this approach in spite of its complexity [5] . However, an example of one of the meteorological methods is presented to show how they can be applied to problems of crop production and water-use efficiency [5] .

The energy balance approach is a suitable example because of its basic nature and its prominence in studies of evapotranspiration. This approach involves determination of the partitioning of energy among the various physical processes contributing to the microclimate, and can be used to measure daily evaporation with little more effort than that required to measure weekly evaporation by soil moisture sampling methods. It has been applied to the determination of evapotranspiration from crops only within the past few years [13] .

4. THE ENERGY BALANCE METHOD

A brief theory of the energy balance is presented here to illustrate what is involved in using such a method. In the total balance, the earth's surface receives more radiation energy than is lost, i.e. the net radiation at the ground surface is positive $-Q_s^*$. The surplus of supplied energy will be transported back to the atmosphere due to two turbulent energy fluxes, the sensible heat flux Q_H and the latent heat flux Q_E, A small portion of the net radiant energy is utilized in the photochemical processes of photosynthesis. A larger portion goes into the heating of the soil, plants, and air. A fairly large portion provides evaporation. Furthermore, energy is transported into the soil due to the ground heat flux Q_G and will be stored by plants, buildings, etc. ΔQ_s. The sensible heat flux is responsible for heating the atmosphere from the surface up to some 100 m during the day, except for days with strong convection. A simplified energy balance equation for a crop community, neglecting the relatively small photosynthetic energy component and assuming horizontal uniformity, can be written as Equatian (2) above.

The terms for sensible heat and latent heat can also be represented by diffusion type equations:

$$Q_H = -\rho C_p K_H \frac{\partial T}{\partial z} \tag{14}$$

$$Q_E = -\rho \lambda K_E \frac{\partial q}{\partial z} \tag{15}$$

where ρ: the air density $(\mathrm{g \cdot cm^{-3}})$; C_p: the specific heat capacity for constant pressure $(\mathrm{calg^{-1}})$; λ: evaporation heat of water; K_H heat transfer coefficient $(\mathrm{cm^2 \cdot s^{-1}})$; K_E water vapor transfer coefficient; z height above ground (cm); and T temperature (˚C).

The ground heat Q_G, is based mainly on molecular heat transfer, and is proportional to the temperature gradient times the thermal molecular conductivity a_G

$$Q_G = a_G \frac{\partial T}{\partial z}. \tag{16}$$

This molecular heat transfer is so weak that during the day only the upper decimeters are heated. When considering the annual cycle of ground temperature, maximum temperature is at the surface during the summer, but 10 - 15 m below the surface during winter [20]. Table 1 shows different values for different thermal molecular conductivity a_G of Equation (16).

Table 1. Thermal molecular conductivity a_G, volumetric heat capacity C_G, and molecular thermal diffusivity v_T for different soil and ground properties [21].

Ground surface	a_G ($W \cdot m^{-1} \cdot K^{-1}$)	C_G (10^6 $W \cdot s \cdot m^{-3} \cdot K^{-1}$)	v_T (10^{-6} $m^2 \cdot s^{-1}$)
rocks (granite)	2.73	2.13	1.28
moist sand (40%)	2.51	2.76	0.91
dry sand	0.30	1.24	0.24
sandy clay (15%)	0.92	2.42	0.38
swamp (90% water)	0.89	3.89	0.23
old snow	0.34	0.84	0.40
new snow	0.02	0.21	0.10

From Equation (16) the determination of Q_G is not practicable because the temperature profile must be extrapolated to the surface to determine the partial derivative there. This can be uncertain because of the high temperature gradients near the surface and the difficulties in the determining thermal heat conductivity. According to [22] using sensitivity analysis, the soil heat flux plate should be buried rather deeply (10 - 20 cm) with some temperature measurements (water protected thermometers) made above it to calculate the heat storage. A similar accuracy can be archived if a temperature profile is used to calculate both the soil heat flux according to Equation (16) at a certain depth and the heat storage between this depth and the surface [19].

The heat exchange in the air due to turbulence is much more effective. This is because turbulent exchange occurs over scales of motions ranging from millimeters to kilometers. Turbulent elements can be thought of as air parcels with largely uniform thermodynamic characteristics. Small-scale turbulence elements join to form larger ones and so on. The largest eddies are atmospheric pressure systems. The heated turbulent elements transport their energy by their random motion. The larger turbulent elements receive their energy from the mean motion, and deliver the energy by a cascade process to smaller elements. Atmospheric turbulence is a specialty of the atmospheric motion consisting in the fact that air volume (much larger than molecules: turbulent elements, turbulent eddy) achieve irregular and stochastic motions around a mean state. They are of different order with characteristic extensions and lifetimes ranging from centimeters and seconds to thousands of kilometers and days [19].

The characteristic distribution of turbulent elements (turbulent eddies) takes place according to their size and is represented by the turbulence spectrum: The turbulence spectrum is a plot of the energy distribution of turbulent elements (turbulent eddies) according to their wavelength or frequency. Depending on the frequency, the distribution is classified as macro, meso or micro turbulence (Foken, 2008). Turbulence refers to the apparently chaotic nature of many flows, which is manifested in the form of irregular, almost random fluctuations in velocity, temperature, and scalar concentrations around their means values in time and space [23] . The calculation of the heat fluxes (sensible and latent) caused by turbulent elements is analogous to Equation (16) using the vertical gradients of temperature T and specific humidity q respectively. The sensible heat flux, Q_H, describes the turbulent transport of heat from and to the earth's surface.

Turbulent transfer is a bulk transfer, so the transfer coefficients K_H and K_E are nearly equal because they are not as much a function of molecular properties as are molecular diffusion coefficients. Substitution of Equations (14) and (15) for the sensible and latent heat terms of Equation (2) and solving for the latent heat term gives:

$$Q_E = \frac{-\left(Q_s^{*} + Q_G\right)}{1 + \frac{C_p}{\lambda}\left(\frac{\partial T/\partial z}{\partial q/\partial z}\right)} \tag{17}$$

Therefore, if we measure the net radiation, soil heat storage, and the air temperature and humidity gradients all of which we are presently able to do, then we can calculate the latent heat flux and water vapor flux which gives the evaporation rate.

5. THE APPLICATION OF METEOROLOGICAL METHODS

The energy balance method has much potential in the study of rate of evaporative loss of water from cropped surfaces because of its fundamental nature and because it can be employed in various other forms. It lends itself to estimating evapotranspiration by approximation procedures when complete data are not available. Such approaches have been suggested and successfully used by [24].

The meteorological methods serve as tools for increasing crop production and water-use efficiency by providing a means of obtaining fundamental information about the controlling physical processes involved in the microclimate.

The economic use of water, for example, requires knowledge of the water evaporated by various vegetated or treated surfaces. It requires an understanding of the effects of stage of growth, plant characteristics, and various management practices on water use and crop production. An understanding of plant response to the microclimate in the evaporation of water is necessary for predicting the amount of water that will be evaporated under given conditions and for controlling evaporation [5].

Furthermore, the meteorological method has a potential for more direct application to these problems. It was suggested several decades ago that the careful determination of evaportranspiration by such a method could be used to determine the time to irrigate [25] (Van Bavel, 1956). This has been tested on an experimental basis and found to be a usable technique. With more information about crop factors and their relationship to water use, the utility of this approach will be increased [5].

As a further development along this line, the scheduling of irrigation of several crops on farms in this area is now being cooperatively carried out on a trial basis. The scheduling is based on meteorological parameters and previously obtained information about the response of crops to certain conditions [5].

This procedure offers an alternative to present methods of determining when to irrigate. The method most commonly used is that of judging irrigation needs by the plant or soil appearance, which often results either in irrigation after the plant has already suffered from lack of water or in the waste of water by irrigating before it is required. Another method is that of using soil moisture measuring instrumentation or sampling procedures to monitor the water content of the soil. These methods can be used satisfactorily on a research basis or on large- scale operations where individuals highly skilled in the interpretation of these instruments are available. However, it is not likely that these instruments will ever be used extensively on the small farming scale. It is highly probable that within the near future the meteorological approach will find widespread application and that with further developments it may be possible to forecast irrigation requirements from crop data, meteorological data, and predicted weather patterns.

Meteorological methods may also be used to study the effects of irrigation on the microclimate and the plant response to these effects. Irrigation decreases air and soil temperatures and increases the humidity of the atmosphere near the surface. It, therefore, exerts a profound influence upon the disposition of energy at the surface and upon the climate near the ground. Such side effects of irrigation may have great importance in the management of suitable microclimates [5] .

Another interesting application is that of studying the modification of the microclimate by such management practices as planting of windbreaks. In another application, the meteorological approach has been used to gain valuable information about the distribution of photosynthesis within a crop throughout the day [26] [27] . Information gained from these studies could lead to the development of an ideal model of a plant so far as leaf arrangement, leaf characteristics and other plant morphological characteristics are concerned. This might help plant breeders in the development of improved plants [5] .

6. CONCLUSIONS

The probable advent of increasing water scarcity in this century will see less increase in irrigated land available for food production than in the past. Novel irrigation technologies need to be tested under local environments and particular agricultural production systems of developing countries like Nigeria. While irrigation can benefit yields and enhance in water-limited environments, the potential for full irrigation is decreasing, with increased competition from the domestic and industrial sectors. Thus the main challenge confronting both rain fed and irrigated agriculture is to improve sustainable water use for agriculture. This can be achieved through 1) an increase in crop water productivity (an increased in marketable crop yield per unit of water taken up by crop), 2) a decrease in water outflows from the plant rooting zone other than that required by plants and 3) an increase in soil water storage within the plant rooting zone through better soil and water management practices at farm and catchment scales.

The micrometeological method of studying the microclimate and all its important physical processes is serving as a useful tool for gaining information about these processes. With this information, we will be better able to predict the response of plants as we alter the natural environmental conditions in an attempt to provide an ideal environment for plant growth. Also, the micrometeorological method may in the near future serve as the basic tool by which we manage our crops and decide when to irrigate our agricultural lands (Wright and Lemon, 1967).

REFERENCES

1. Thornthwaite, C.W. (1948) An Approach toward a Rational Classification of Climate. Geographical Review, 38, 55-94.
2. Tucci, C.E.M. and Beltrame, L.F.S. (2009) Evaporaç&atildeo e evapotranspiraç&atildeo, In: Tucci, C.E.M., Ed., Hidrologia: Ciência e Aplicaç&atildeo, UFRGS Editora, Porto Alegre, 253-287.
3. Allen, G.R., Pereira, S.L., Raes, D. and Smith, M. (1998) Crop Evapotranspiration: Guidelines for Computing Crop Water Requirements. Food and Agrocultural Organization of the United Nations, FAO-56, Rome, 300 p.
4. Allen, R.G., Pereira, L.S., Howell, T.A. and Jensen, M.E. (2011) Evapotranspiration Information Reporting: I. Factors Governing Measurement Accuracy. Agricultural Water Management, 98, 899-920.
5. Wright, J.L. and Lemon, E.R. (1967) Micrometeorological Methods as Tools for Increasing Crop Production and Water-Use Efficiency. Proceedings of 8th Annual Fertilizer Conference of the Pacific Northwest, Twin Falls, 11-13 July 1967.
6. Munn, R.E. and Davidson, B. (1957) Exploring the Atmosphere's. Pergamon Press, New York, 376 p.
7. Bezerra, J.R.C., de Azevedo, P.V., da Silva, B.B. and Dias, J.M. (2010) Evapotranspiraç&atildeo e coeficiente de cultivo do algodoeiro BRS-200 Marrom, Irrigado. Revista Brasileira de Engenharia Agrícola e Ambiental, 14, 625-632.
8. Gavilán, P. and Berengena, J. (2007) Accuracy of the Bowen Ratio-Energy Balance Method for Measuring Latent Heat Flux in a Semiarid Advective Environment. Irrigation Science, 25, 127-140.
9. Hou, L.G., Xiao, H.L., Si, J.H., Zhou, M.X. and Yang, Y.G. (2010) Evapotranspiration and Crop Coefficient of Populus euphratica Olivi Forest during the Growin Season in the Extreme arid Region Northwest China. Agricultural Water Management, 97, 351-356.
10. Steduto, P. and Hsiao, T.C. (1998) Maize Canopies under Two Soil Water Regimes: IV. Validity of Bowen Ratio-Energy Balance Technique for Measuring Water Vapor and Carbon Dioxide Fluxes at 5-min Intervals. Agricultural and Forest Meteorology, 89, 215-228.

11. Todd, R.W., Evett, S.R. and Howell, T.A. (2000) The Bowen Ratio-Energy Balance Method for Estimating Latent Heat Flux of Irrigated Alfalfa Evaluated in a Semi-Arid, Advective Environment. Agricultural and Forest Meteorology, 103, 335-348.
12. Sergeant, D.H. and Tanner, C.B. (1967) A Simple Psychrometric Apparatus for Bowen Ration Determinations. Journal of Applied Meteorology, 6, 414-418.
13. Tanner, C.B. (1960) Energy Balance Approach to Evapotranspiration from Crops. Soil Science Society of America Journal, 24, 1-9.
14. Liu, H. and Foken, T. (2001) A Modified Bowen Ratio Method to Determine Sensible and Latent Heat Fluxes. Meteorologische Zeitschrift, 10, 71-80.
15. Geiger, Z.R. (1959) The Climate near the Ground. 2nd Edition, Harvard University Press, Cambridge, MA, 482 p.
16. Slatyer, R.O. and McIlroy, I.C. (1961) Practical Microclimatology. UNESCO, Prepared and Reproduced by CSIRO, Australia.
17. Waggoner, P.E., Ed. (1965) Agricultural Meteorology. American Meteorological Society, Boston, 188 p.
18. Evans, L.T. (1962) Environmental Control of Plant Growth. Academic Press, New York, 449 p.
19. Foken, T. (2008) Micrometeorology. Springer-Verlag, Berlin, 306 p.
20. Lehmann, A. and Kalb, M. (1993) 100 Jahre meteorologische Beobachtungen an der S?kularstation Potsdam 1893-1992. Deutscher Wetterdienst, Offenbach, 32 p.
21. Stull, R.B. (1988) An Introduction to Boundary Layer Meteorology. Kluwer Academic Publishers, Dordrecht, Boston and London, 666 p.
22. Liebethal, C. and Foken, T. (2007) Evaluation of Six Parameterization Approaches for the Ground Heat Flux. Theoretical and Applied Climatology, 88, 43-56.
23. Arya, S.P. (2001) Introduction to Micrometeorology. Academic Press, San Diego, 415 p.
24. Jensen, M.E. and Haise, H.R. (1963) Estimating Evapotranspiration from Solar Radiation. Journal of the Irrigation and Drainage Division, 89, 15-41.
25. Van Bavel, C.H.M. (1956) Evapotranspiration, the ET Method, a New Way to Find a Right Time to Irrigate. Better Farming Methods, April 1956, 14.

26. Wright, J.L. and Lemon, E.R. (1966) Photosynthesis under Field Conditions. VIII. Analysis of Windspeed Fluctuation Data to Evaluate Turbulent Exchange within a Corn Crop. Agronomy Journal, 58, 255-261.

27. Wright, J.L. and Lemon, E.R. (1966) Photosynthesis under Field Conditions. IX. Vertical Distribution of Photosynthesis within a Corn Crop. Agronomy Journal, 58, 265-268.

Chapter 8

Upflow Evapotranspiration System for the Treatment of On-Site Wastewater Effluent

Sean Curneen and Laurence Gill

Department of Civil, Structural and Environmental Engineering, Trinity College Dublin, Dublin 2, Ireland

ABSTRACT

Full-scale willow evapotranspiration systems fed from the base with septic tank or secondary treated domestic effluent from single houses have been constructed and instrumented in Ireland in order to investigate whether the technology could provide a solution to the problem of on-site effluent disposal in areas with low permeability subsoils. Continuous monitoring of rainfall, reference evapotranspiration, effluent flows and water level in the sealed systems revealed varying evapotranspiration rates across the different seasons. No system managed to achieve zero discharge in any year remaining at maximum levels for much of the winter months, indicating some loss of water by lateral exfiltration at the surface. Water sampling and analysis however, showed that the quality of any surface overflow from the systems was similar to rainfall runoff. The performance results have then been used to formulate design guidelines for such systems in Ireland's temperate maritime climate. The effect of varying different combinations of design parameters (plan area, soil depth, etc.) has been evaluated with respect to the simulated number of overflow days over a five-year period using a water balance model. Design guidelines have then been based upon minimising the amount of runoff, in conjunction with other practical and financial considerations.

Keywords: on-site wastewater; low permeability; evapotranspiration; willow; passive treatment

1. INTRODUCTION

The domestic wastewater of approximately one third of the population in Ireland (~500,000 dwellings) is treated on-site by domestic wastewater treatment systems (DWWTS) of which more than 87% are septic tanks [1]. If situated and constructed incorrectly, the potential impacts of such on-site effluent include the pollution of either groundwater and/or surface water. In particular, areas with inadequate percolation due to low-permeability subsoils and/or insufficient attenuation due to high water tables and shallow subsoils present the greatest challenge in Ireland for dealing with effluent from DWWTS. If there is insufficient permeability in the subsoil to take the effluent load, ponding and breakout of untreated or partially treated effluent at the surface may occur with associated serious health risks. There will also be a risk of effluent discharge/runoff of pollutants to surface water and to wells which lack proper headwork or sanitary grout seals [2]. It is estimated that the overall proportion of the country with inadequate conditions for DWWTS either all year round or that can arise intermittently during wet weather conditions is 39% [3].

The specification [4] of a lower limit to subsoil permeability (defined according to an on-site falling head percolation test) for effluent discharge to ground, in conjunction with the fact that surface water discharges are generally not being licensed for one-off housing, means that many areas will be deemed unsuitable for single house development. To address these problematic areas and allow development while protecting water resources from the risk of pollution by on-site effluent, alternative wastewater treatment and disposal options are needed. One option that has been considered is the concept of discharging the on-site wastewater effluent into a sealed basin and relying on the net evapotranspiration (ET) from willow trees to exceed the rainfall and effluent hydraulic loads, thereby creating a zero-discharge treatment solution. This technology has been introduced into Denmark for on-site wastewater treatment with some success [5,6] with national guidelines produced [7]. Willows are highly suitable for such an application in such temperate climates due to their high transpiration rates throughout the growing season [8,9], efficient uptake of nutrients [10,11], tolerance of flooded soils and oxygen shortage in the root zone [12] and resilience to pollutants [13]. In more tropical climates other species have been proven to perform such a role; for example, trials in self-recirculating systems in Australia have demonstrated zero discharge using mixed species of bamboo and citrus fruits [14].

Hence, a series of full-scale trials was established in Ireland to evaluate the use of sealed basin ET systems to treat on-site wastewater effluent in Ireland.

These systems have been monitored over a five-year period and the results obtained from these trials have been used to draft national design criteria and to determine the operating limits for these systems in an Irish context.

2. MATERIALS AND METHODS

2.1. Construction of Full-Scale ET Systems

Thirteen full scale willow systems on 10 sites were constructed to treat domestic wastewater effluent from single house dwellings across Ireland: 10 in County Wexford, 1 in County Limerick and 2 in County Leitrim, as shown on Figure 1. The sites were located in areas with very low permeability (i.e., clayey) subsoils, in which effluent percolation would not be a feasible disposal method for the on-site effluent. The systems were designed to specifications that would allow for the long-term study and comparison of key parameters (effluent type, willow species, plan area, aspect ratio and effluent distribution) as detailed in Table 1. The systems were designed (area and depth) based upon a modeled water balance between the typical wastewater effluent rates expected from the dwelling, the local rainfall and the estimated evapotranspiration from the basin over a four year period. Realistic time-varying fluctuations in flows (from basis of previous research carried out by Gill et al. [15]), local meteorological conditions and estimated crop coefficients were input into the water balance model which then simulated willow evapotranspiration across a four year period. The crop coefficients were assumed to be 1 throughout the winter months, increasing up to maximum values of 2.65 across the summer months. At each time step, the calculated stored volume was then divided by the porosity of the soil expected within each willow system (estimated at 30%) to determine the depth fluctuations. Different combinations of area and depth were compared in order to determine the design dimensions of each system which should (theoretically) ensure zero discharge across the 4 year simulation period. The choice of depth was also guided by previous Scandinavian experience with such ET systems using the same willow varieties [7].

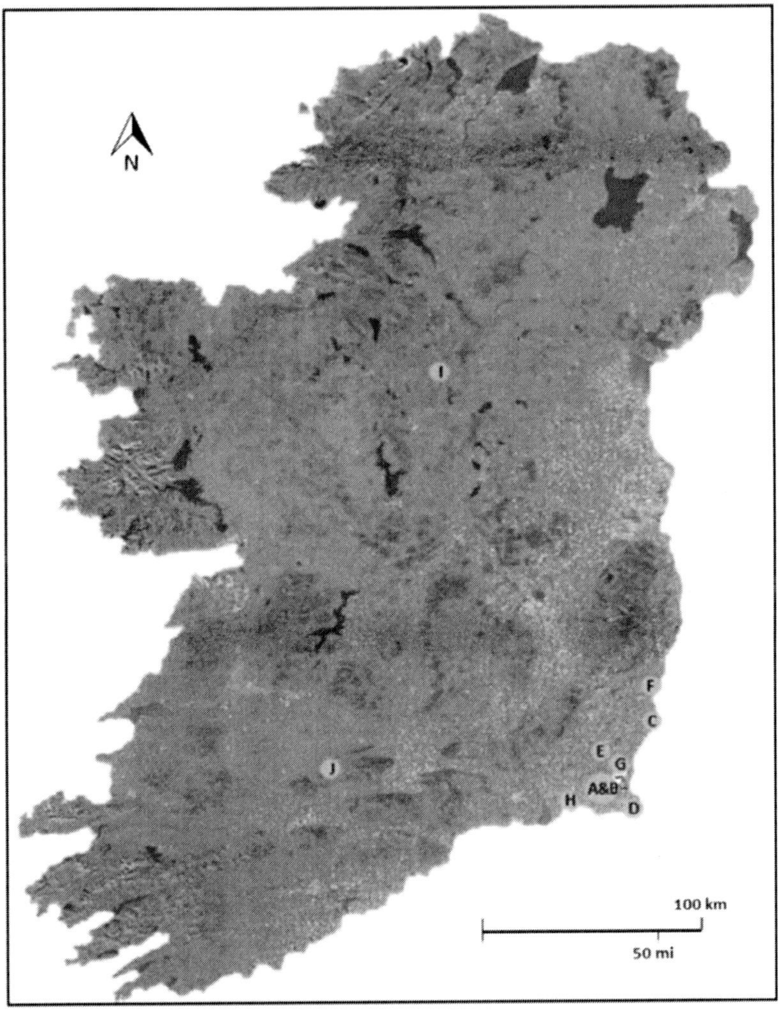

Figure 1. The location of the 13 full-scale ET systems researched in Ireland.

The basins were excavated to a depth of 1.8 m with a geosynthetic barrier laid along the bottom and sides of the excavated basin. An impermeable membrane (minimum 0.5 mm thickness butyl rubber or sealed low-density polyethylene (LDPE) was then laid on top of the geosynthetic barrier followed by a second geosynthetic barrier laid on top of the impermeable membrane. For most systems, the effluent was distributed either by pumping or gravity flow into 110 m rigid diameter pipes at 4 m spacing along the base of the basins bedded in 300 mm depth of 8–32 mm diameter gravel. The excavated soil was then back-filled into the basins up to ground level. On Sites A and B however, the effluent was pumped on a volume-dose basis into a semi-rigid 40 mm diameter distribution pipe at 1 m depth. It should be noted that the excavated soil was used to re-fill the systems due to economic considerations. Importing such a quantity of material would also require more than 50 truckloads of imported earth for a single house which was deemed to be unacceptable for a single house, both logistically and environmentally.

Table 1. Summary of design parameters compared between 13 full-scale systems.

Site	Constructed	Effluent [a]	Area (m^2)	Aspect Ratio (length:width)	Distribution	Willow Variety
A	May 2009	STE	296	4.6:1	pumped	*Bjorn, Tora, Jorr*
B	May 2009	STE	570	1.6:1	pumped	*Bjorn, Tora, Jorr*
C	April 2010	STE	420	2.9:1	pumped	*Tora, Torhild, Tordis*
D	May 2010	STE	464	1.8:1	gravity	*Tora, Torhild, Olof*
E	April 2010	STE	24	2.7:1	gravity	native Irish species
F(1)	July 2010	SE	900	1.4:1	gravity	*Tora, Tordis, Olof*
F(2)	July 2010	SE	900	9.0:1	gravity	*Tora, Tordis, Olof*
G	September 2010	SE	340	3.4:1	gravity	*Tora, Tordis, Olof*
H(1)	April 2011	STE	260	4.1:1	pumped	*Tora, Tordis, Inger*
H(2)	April 2011	STE	260	4.1:1	pumped	*Tora, Tordis, Inger*
I(1)	May 2011	SE	216	3.4:1	gravity	*Tora, Torhild, Olof*
I(2)	May 2011	SE	216	3.4:1	gravity	*Tora, Tordis, Torhild*
J	May 2012	STE	170	6.8:1	pumped	*Tora, Torhild, Tordis*

Notes: [a] STE—septic tank effluent; SE—secondary treated effluent.

Willow cuttings (see Table 1 for varieties) were planted at a density of 3 per m^2. Coppicing of all the willows was then carried out, in accordance with standard short rotation coppice guidelines [16,17] whereby all shoots were cut back as close to the ground as practically possible in the first dormant season to encourage vigorous growth and multiple shoot development in the following season. Cutback was carried out in a three year cycle thereafter, with alternating thirds of the willow system cut every year.

2.2. Instrumentation

Each ET system was designed in order to quantify the hydraulic loading via rainfall and effluent from which a water balance calculation was used to determine the removal of water from the system via evapotranspiration. A tipping bucket device was used to measure flow continuously on the sites that had a sufficient slope for a gravity feed from the proprietary system to the willow system (as well as splitting the flows evenly between systems on sites with two willow systems). A reed switch on the tipping bucket was connected to a Solinist® Rainlogger Model 3002 (Georgetown, ON, Canada) datalogger. On sites where pumping of effluent to the willow system was required, OTT

Thalimedes (Kempten, Germany) water level meters (accuracy: ±0.002 m) were installed in the sump to measure continually the water level. The water level within each ET system was measured on an hourly basis using an OTT Orpheus Mini groundwater datalogger (accuracy: ±0.05%). Campbell Scientific (Loughborough, UK) weather stations were installed to record temperature, relative humidity, wind speed, solar radiation and net radiation. Campbell Scientific ARG100 tipping bucket rain gauges were installed on all sites and either connected to the Campbell Scientific dataloggers or to Solinist® Rainlogger (Model 3002) dataloggers.

2.3. Water Quality Analysis

All septic tank effluent (STE), secondary treated effluent (SE), samples of stored water within each system (taken from the inspection wells) and ponded surface water were periodically analysed for chemical oxygen demand (COD), total nitrogen (TN), ammonium (NH_4^+–N), nitrite (NO_2^-–N), nitrate (NO_3^-–N), orthophosphate (PO_4^{3-}–P), chloride (Cl^-) and sulphate (SO_4^{2-}) using a Merck Spectroquant Nova 60® spectrophotometer (Darmstadt, Germany). In addition, indicator bacteria of faecal contamination, Total Coliforms (TC) and E. coli were analysed for using the IDEXX Colilert®-18 test (Westbrook, MA, USA) with enumeration carried out using IDEXX Quanti-Tray®/2000, a semi-automated quantification method based on the Standard Methods Most Probable Number (MPN) model.

2.4. Soil Porosity

In order to calculate the water balance accurately, the available pore space in the soil within the willow ET system needed to be calculated from which the depth of effluent in the system could be converted into a stored volume. Replicate 200 mm deep soil samples were collected in 100 mm diameter steel cores at a variety of depths from within the ET systems on each site at the end of the growing season when the water levels within the willow systems were low. The samples were transported back to the laboratory and extruded and then weighed immediately to represent the weight of a fully drained (in situ) sample. The sample length was then submerged in water until saturation point was reached. Upon removal, the soil sample was then reweighed to determine the water content at the saturation point. In addition, the resultant change in water level within each ET system in response to single peak, short duration, large rainfall events were analysed from the field data. Such short rainfall events were used to minimise the effect that any ongoing evapotranspiration would have on the water level. Knowing the net rainfall falling onto the system and respective rise in water level, the effective porosity (usable storage volume) could be calculated and then averaged out for a series of events. It should be noted that the effective porosity of any gravel in the system was also included in the depth-volume calculations when appropriate, using a measured gravel porosity of 32%.

3. RESULTS FROM THE FIELD TRIALS

3.1. Loadings

A summary of the data collected at each site is shown in Table 2 with examples of rainfall and potential ET (ET_o) from 2013. A more detailed discussion of the field results is discussed in Curneen and Gill [18].

Table 2. Summary hydraulic loading data from ET system trials.

Site	Effluent Production Started	Population Equivalent (PE)	Mean Daily Flow (L/d)	Effective Soil Porosity (%)	Annual Rainfall (2013) (mm)	Annual ET_o (2013) (mm)
A	October 2010	4	628.4 [a]	15.1	981	585
B	January 2012	1	443.1 [a]	15.6	981	585
C	December 2010	2	356.7	11.0	933	584
D	May 2010	4	483.3	11.0	847	592
E	April 2011	n/a [b]	3.0	21.0	833	571
F(1)	January 2013	3	306.6	18.0	811	587
F(2)	January 2013	3	306.6	19.0	811	587
G	n/a [b]	0	0	9.6	928	610
H(1)	March 2012	2	181.3	9.5	930	602
H(2)	March 2012	2	181.3	9.5	930	602
I(1)	November 2012	2	163.4	16.0	1158	465
I(2)	November 2012	2	163.4	20.0	1158	465
J	November 2012	3	330.3	19.0	1039	526

Notes: [a] suspected rainwater infiltration into effluent; [b] an ET system was constructed for a new house but no-one moved into the house during the monitoring period.

3.2. Water Level Fluctuations

Although all the willow ET systems had been designed to operate as zero discharge systems on the basis of a theoretical water balance, the water level data (as well as observations at the sites) revealed that no system managed to achieve zero discharge in any year, remaining full for much of the winter months. This indicated ongoing loss of water by lateral exfiltration at the surface during these winter periods. Overflows did also occur to a certain extent during the summer of 2012 when precipitation levels were exceptionally high.

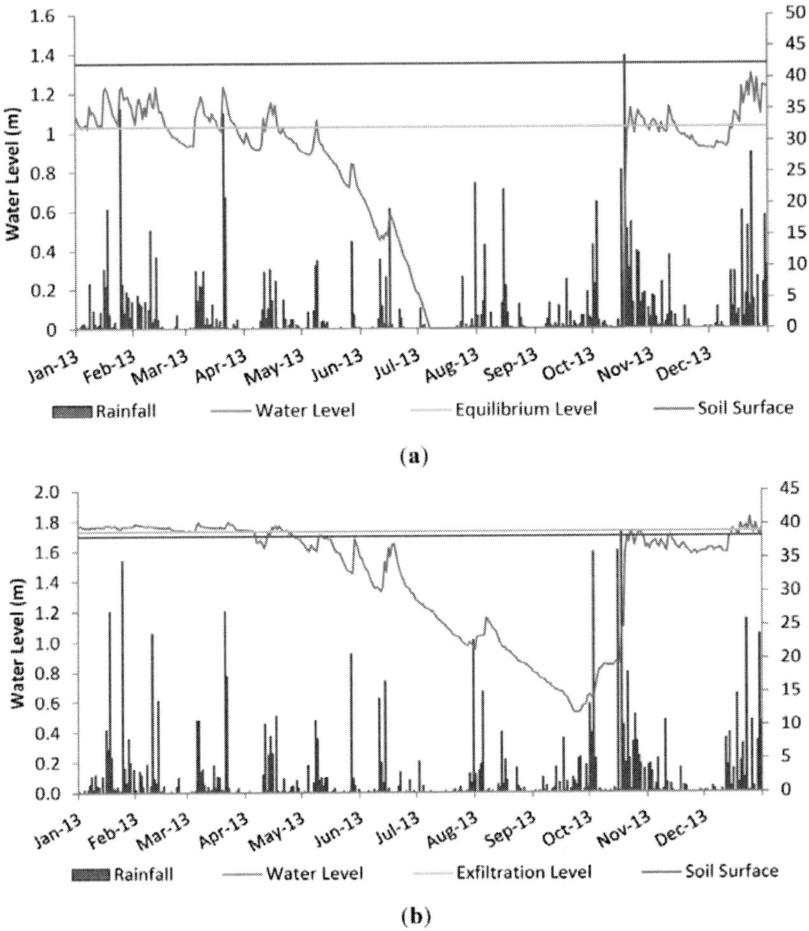

Figure 2. Water level and daily rainfall (2013) for (**a**) Site A; (**b**) Site C.

An example of annual water level profiles within two willow systems is shown in Figure 2. The equilibrium level (the level above which overflow from the system occurs) along with the soil surface level are illustrated, both of which are measured relative to the base of the system. For example, at Site A (Figure 2a) overflow was determined to be occurring at a depth of 1.02 m even though the soil surface level was 1.35 m, suggesting that either the impermeable liner did not reach up to the top of the system or it had been breached at this depth. It can also be observed that the system was empty for most of the summer months, meaning that optimal evapotranspiration performance was probably not achieved during that period. This did not occur at Site C (Figure 2b), primarily due to the system being much deeper with an overflow level at 1.73 m from the base. It is also interesting to note the effect of large October rainfall events on the water level in the systems, in response to which both systems refilled to capacity in a very short space of time, which is indicative of the low effective soil porosities.

3.3. Water Balance (ET and Crop Factors)

The actual evapotranspiration from each system throughout the monitoring period was determined on the basis of a water balance taking into account the measured hydraulic loadings from rainfall and wastewater, as shown in Equation (1):

$$\mathrm{ET_{wilow}} = R + I + (\Delta H \times V_u \times A) \tag{1}$$

where, ET_{willow} = evapotranspiration from willow system; R = rainfall; I = influent (wastewater); ΔH = change in water level within system; V_u = usable water volume storage within soil/gravel (i.e., effective porosity); A = surface area.

Evapotranspiration rates attained by the willow systems were calculated on a daily basis with the results then averaged and presented on a monthly basis. At such times when the water level was above the overflow point, ET_{willow} could not be accurately calculated using water balance Equation (1), as the volume of water leaving the system due to any exfiltration/overflow could not be quantified. Hence, for such flooded periods during the growing season, mean daily ET rates for those days were assumed to be the same as the closest days during which ET_{willow} rates could be rigorously calculated using Equation (1). In the winter months when the trees were dormant, it was assumed that ET_{willow} was equal to ET_o on days where the water level was above the overflow level: on all other days, the actual ET_{willow} was calculated according to Equation (1). Monthly crop coefficients (K_c) were calculated by dividing the assumed monthly ET_{willow} by the ET_o for the month.

The total assumed ET_{willow} for the monitored period in each calendar year for each willow system shown in Table 3 reveals significant variation between the willow systems, even for those systems in which the willow cuttings had been planted in the same year. Several reasons are thought to attribute to this including different effluent loadings, different soil porosities and different rates of establishment/development of the willow trees. The highest ET_{willow} rates were calculated for Sites A, B, C, and F(2), which was expected as these sites undoubtedly had better willow tree development (also aided by effluent loading from the outset for Sites A and C).

Table 3. Annual calculated ET_{willow} for each willow system.

System	2010		2011		2012		2013		2014	
	ET_{willow}	months	ET_{willow}	months	ET_{willow}	months	ET_{willow}	months	ET_{willow}	months
A	755	10	929	12	725	12	928	12	-	-
B	554	7	638	12	765	12	723	12	-	-
C	358	4	717	12	670	12	752	12	845	12
D	-	-	558	12	495	12	487	12	-	-
E	282	7	573	12	669	12	90	4	-	-
F(1)	-	-	511	10	595	12	577	12	598	12
F(2)	-	-	641	10	827	12	817	12	930	12
G	-	-	341	7	574	12	445	8	-	-
H(1)	-	-	-	-	864	9	788	12	-	-
H(2)	-	-	-	-	771	9	576	12	-	-
I(1)	-	-	-	-	424	9	613	12	-	-
I(2)	-	-	-	-	402	9	540	12	-	-
J	-	-	-	-	-	-	680	12	-	-

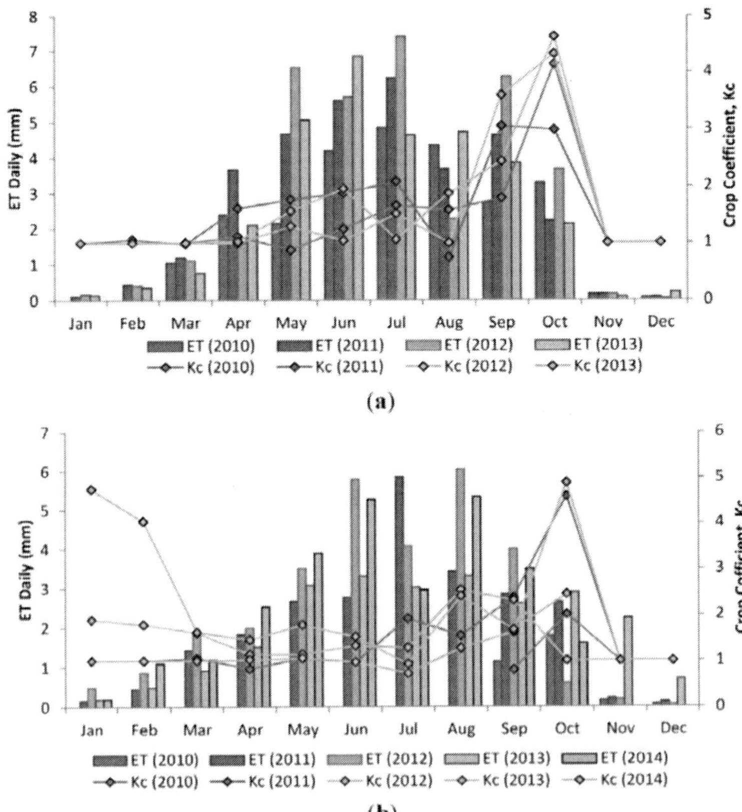

Figure 3. Monthly average ET_{willow} rates for (**a**) Site A and (**b**) Site C.

Typical ET_{willow} results and crop coefficients from two of the longest operating systems (Site A and Site C) are shown in Figure 3. As expected, the highest ET_{willow} rates were attained in the summer months during the willow trees' growing season. It might have been expected that the ET_{willow} rates would increase year on year, as the willow trees developed and became larger. However, the mean daily ET_{willow} for each month on Site A shows that this was not always the case. While the mean daily ET_{willow} did increase year on year from 2010 to 2012 for the months May, June, July and September, it decreased for all these months (except June) in 2013. This is probably due to the low water availability within the system at these periods, thus preventing the willows from being able to attain their optimum potential evapotranspiration, and possibly an indication that the roots had not developed down to the base of the system.

A comparison of the crop coefficients also shows a similar result. Most of the monthly ET_{willow} highs (as well as associated crop coefficients) throughout the growing season occurred in 2012 when water availability was at its highest due to the high rainfall levels, not in 2013 when the willow trees were more developed and reference evapotranspiration rates were higher. It is also interesting to note the relatively high crop coefficients in September and October, which shows that there was still a significant amount of water being evapotranspired even though the ET_o values were relatively low.

3.4. Water Quality

The quality of the effluent discharged into the systems, the water in each system's monitoring well and ponded water on the surface during winter is shown in Table 4. An interesting comparison can be made between the water quality in the sumps from those systems that were receiving effluent and those monitored before effluent was introduced. As would be expected, the nutrient concentrations were marginally higher in the samples from the systems receiving effluent compared to the samples from the systems with no effluent. However, the results show that the willow systems acted as an excellent pollutant attenuation process on the influent, promoting greatly reduced organic, nutrient and indicator bacteria concentrations. Equally, the quality of standing water during winter periods on top of the systems receiving effluent proved similar to the sump samples from the systems that had not received effluent, with the exception of very low levels of indicator bacteria.

Table 4. Mean wastewater influent, internal sump water quality and ponded surface water.

water quality parameter	STE	ET Systems Receiving Effluent		ET Systems with No Effluent
		ET System Sump	Ponded Water on Surface	ET System Sump
COD (mg/L)	377 ± 96	42.6 ± 8.9	29.8 ± 6.7	32.9 ± 7.1
Total-N (mg/L)	50.3 ± 8.7	9.8 ± 2	12.4 ± 4.1	8.1 ± 1.4
Ortho-P (mg/L)	6.3 ± 1.2	0.55 ± 0.4	0.24 ± 0.14	0.30 ± 0.08
E. coli no./100 mL *	9.2×10^5	22.1	8.3	<1

Note: * geometric mean

4. EVALUATION OF DESIGN PARAMETERS ON PERFORMANCE

4.1. Model to Simulate ET System Performance

In order to use the results of the field trials to develop appropriate national design guidelines, modelling was first carried out to predict the effect of varying the different design parameters using realistic crop factors and meteorological data as determined on site from the field trials. The design model for sizing of a willow system was based on a daily water balance (i.e., Equation (1)) which incorporated hydraulic loading into the system via rainfall and effluent and removal of water from the system via evapotranspiration. Daily rainfall and reference evapotranspiration measured on Sites A and B over the duration of the monitoring period was used as inputs into the model to produce the resultant daily water level profiles over a five-year monitored period (January 2010–December 2014).

The usable effective porosities used in the model varied according to the medium at and below any particular water level, as shown schematically on Figure 4: changes in water level occurring within the soil medium assumed a V_u value of 15%; changes within the gravel layer used a V_u of 33%; and changes in water level above the surface of the soil assumed V_u to be 100%.

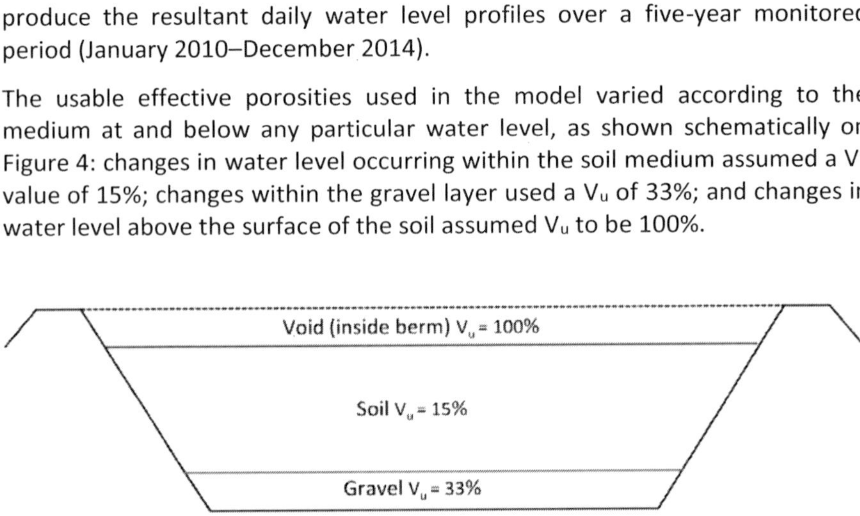

Figure 4. Sample cross-section of ET system with relevant effective porosities (V_u).

The ET rates and crop coefficients (K_c) used were based on the results from the systems at Sites A, B, C and F(2) which were deemed to have performed to the expected standard over the duration of the research project as they had not been compromised by poor growth or low effluent loading as many of the other sites had. For Sites A and B, it was considered that conservatively low crop coefficients were calculated across some of the summer months due to the fact that the systems had emptied out quite early in two of the monitored growing seasons, leaving little water readily available for evapotranspiration. This is illustrated in Figure 5 which shows a comparison between the crop coefficients for the two systems with those determined in irrigated lysimeter experiments carried out by Guidi et al. [19] for willows in the first and second growing seasons: the crop coefficients on Sites A and B were equal to or higher than the lysimeter experiments for May and June but then dropped back for the rest of the summer months when the systems were empty. As such, it was deemed that these crop coefficients were lower than values that could be expected had water been more readily available if the systems had been deeper. Hence, the representative K_c values used for the model shown in Table 5 were chosen based the crop coefficients determined in the field trials but also influenced by the higher values determined by Guidi et al. [19] in August. The dormant season crop coefficients were determined by taking the average values from all systems as the results showed that willow development appeared to have a minimal bearing on these values. The annual crop coefficient used to model the ET systems works out at 2.2 which, for a mean annual ET_o on Sites A and B of 517 mm across the monitoring period, would give an expected annual total of 1135 mm of evapotranspiration from the modelled system.

Table 5. Representative K_c values and resulting system ET rates in the model.

Month	Assumed K_c	Mean ET_o (mm)	ET_{willow} (mm)
January	1.5	4.4	9.1
February	1.4	11.0	20.8
March	1.4	32.1	48.0
April	1.3	59.2	77.0
May	1.5	82.2	123.4
June	1.9	92.1	175.0
July	2.1	84.4	177.2
August	3.4	76.7	260.9
September	3.6	46.7	168.1
October	3.2	20.9	66.8
November	1.4	4.9	6.8
December	1.2	1.9	2.3
Annual Mean	2.2	516.6	1135.4

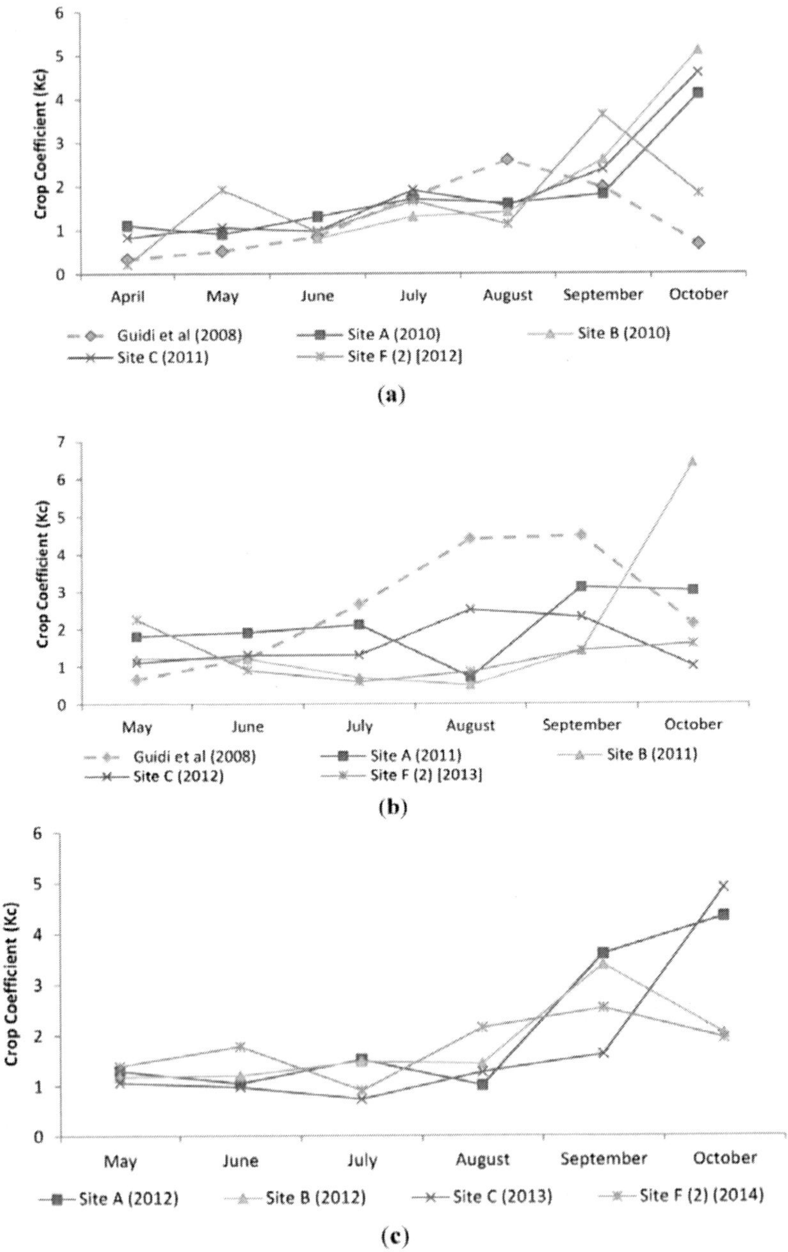

Figure 5. Crop coefficients for Sites A, B, C and F(2) compared to those determined by Guidi et al. [18] for (**a**) first growing season; (**b**) second growing season and (**c**) third growing season.

4.2. Comparison of Design Parameters

The ET_{willow} water balance model was then used to predict the water level profile response of the system to various different physical design factors: the plan area of system, the effective porosity of the storage medium (soil and gravel), the depth of soil within the willow system basin, the depth of gravel layer within the system and the depth of free space available above soil surface (maintained by a raised boundary (bund)). For each simulation, the five years of rainfall and reference evapotranspiration data from the field trials (2010 to 2015) were used as inputs. The model then simulated the expected water level profile and any predicted overflow for any defined wastewater loading. Various combinations (within practical constraints) of the above factors were used in order to determine the minimum area required to try to reduce any runoff from the systems as much as possible throughout the year. The model showed that the evapotranspiration that could be expected from a willow ET system in a given year in Ireland's climate would generally be lower than the corresponding hydraulic load (rainfall and effluent), resulting in overflow from the system occurring, as was the case in 2012. This mirrored the findings on most of the systems monitored. Hence, simulations were carried out to seek to optimise the system in terms of lowering the amount of overflow days that may occur, while also taking practical area and financial constraints into account.

4.2.1. Plan Area and Overall Depth

The effect of plan area and overall system depth on water level depth was examined using simulations based on a dwelling with a population equivalent (PE) of 4 with the other design parameters listed in Table 6, fixed. The effect of increasing the area on the number of overflow days is illustrated in Figure 6 which shows that significant gains with respect to decreasing the amount of overflow days can be made by increasing the plan area of the ET system from 250 up to 500 m^2. However, any further gains made by increasing the surface area beyond 500 m^2 appear to be marginal and would probably not justify the significant additional construction costs required.

An example of a water level profile produced by the model for a 500 m^2 plan area system for a 4 PE residence is shown in Figure 7. The system was determined to have overflow for 141 days across the five-year period corresponding to the monitoring of the full-scale systems. It is worth noting that water level would be predicted to reach the bottom of the system by the end of August in four out of five of the growing seasons monitored.

Table 6. Fixed parameters used for comparative model analyses of system design.

Area (m^2)	Effluent (L/c.day)*	Soil Porosity (%)	Gravel Porosity (%)	Depth of Gravel Layer (m)	Depth to Soil Surface (m)	Depth of Free Space (m)
500	100	15	35	0.3	1.8	0.3

Note: * litres per capita per day

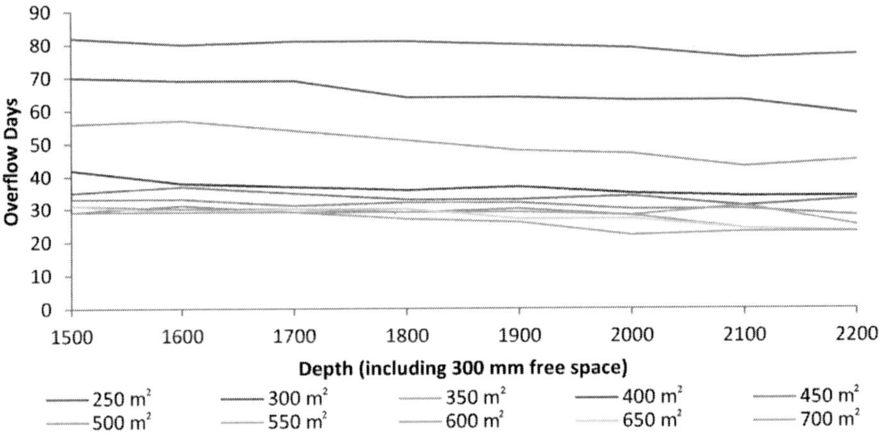

Figure 6. Effect of plan area on the mean number of overflow days simulated per year.

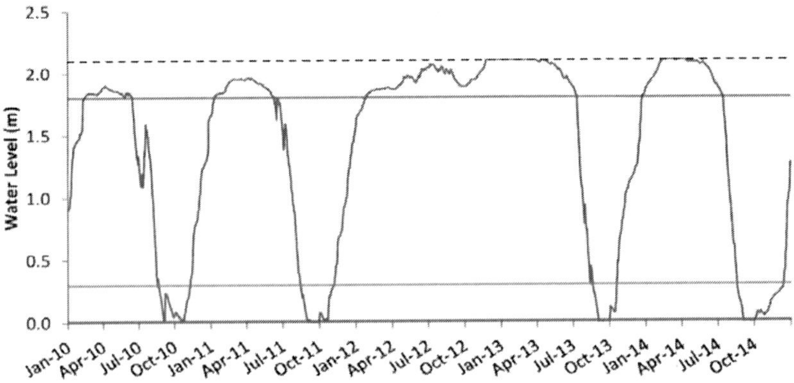

Figure 7. Water level profile simulated by the water balance model for a 500 m^2 ET system receiving effluent from a 4 PE house.

The effect of increasing the overall depth on the amount of overflow from the willow system in Figure 8 shows that it would be beneficial for smaller areas (<350 m^2) and to a lesser extent for larger areas (>550 m^2). For the range of areas in between, however, it appears that not much benefit can be gained by increasing the depth of the system.

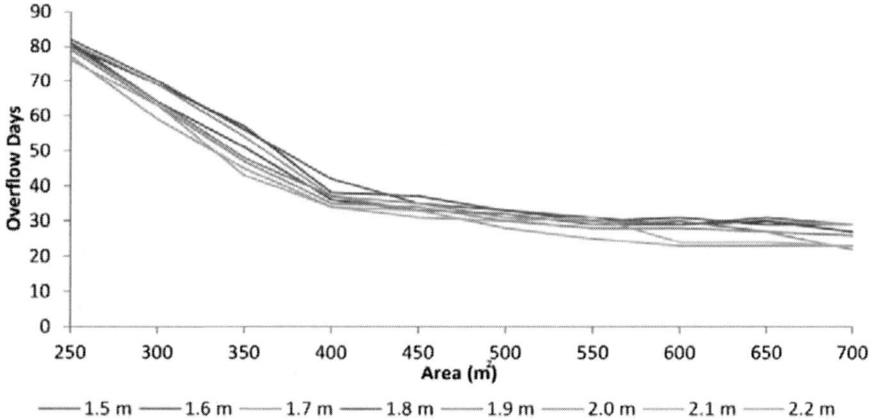

Figure 8. Effect of increasing the depth on the mean number of overflow days simulated per year.

A matrix highlighting the predicted overflows for different combinations of area and depth is illustrated in Table 7. As would be expected, the combinations of higher area and depth result in lower predicted days with overflow. Several combinations result in approximately 30 to 35 days of overflow per year, which is probably the realistic optimal minimum, taking practical and financial constraints into account. The effect on the water level profile of increasing the depth of a 500 m^2 willow system from 1.5 to 1.8 m, for example, has a limited effect and the number of overflow days, decreasing from 35 to 33. A further increase in depth of the 500 m^2 area system to 2.1 m again has limited effect, with a predicted decrease of just 1 overflow day per year over the duration of the modelled period. The overflow days can be reduced to as low as 22 days per year for a 4 PE dwelling, but this would require a significantly larger system of area 700 m^2 and overall depth of 2.0 m, at which point such ET systems for single houses would not be deemed to be financially viable in an Irish context.

The result of running the model again for four different effluent loadings is shown in Figure 9 and Table 8. The loading rates were 5, 6, 8, and 10 PE, while assuming a daily effluent production of 100 L/c.day. While this effluent loading rate is lower than that assumed in the Irish Code of Practice [4], it is based on the assumption that any dwelling using a willow ET treatment system should have up-to-date water saving appliances (e.g., dual flush toilet, low-flow shower head, tap aerators) whereby such a wastewater production would be realistic.

Table 7. Matrix showing the mean number of overflow days simulated per year from the willow system for different area and depth combinations.

Area	Depth of ET System							
	1.5 m	1.6 m	1.7 m	1.8 m	1.9 m	2.0 m	2.1 m	2.2 m
250 m^2	82	80	81	81	80	79	76	77
300 m^2	70	69	69	64	64	63	63	59
350 m^2	56	57	54	51	48	47	43	45
400 m^2	42	38	37	36	37	35	34	34
450 m^2	35	37	35	33	33	34	31	33
500 m^2	33	33	31	32	32	30	30	28
550 m^2	31	30	30	29	29	28	31	25
600 m^2	29	31	29	29	30	28	24	23
650 m^2	31	29	30	30	27	27	24	23
700 m^2	29	29	29	27	26	22	22	22

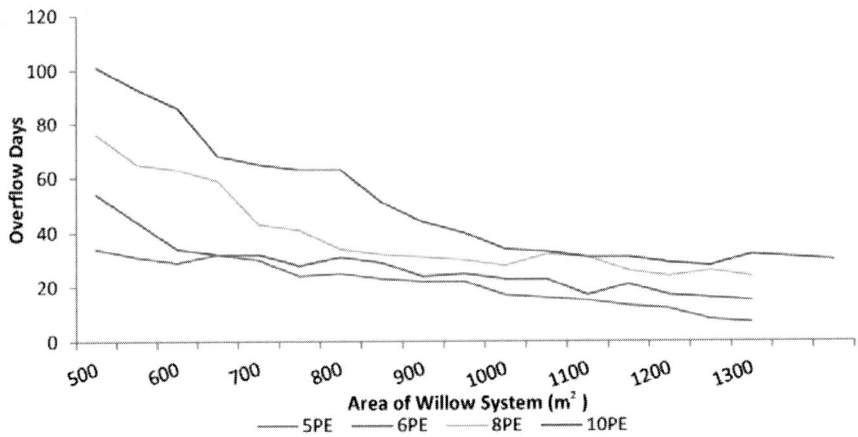

Figure 9. Effect of increasing the plan area of the ET system on the mean number of overflow days simulated per year for different hydraulic loading scenarios.

This analysis shows that reductions in the number of overflow days plateau off once a certain area is reached. Lowering the number of overflow days below the 30–35 range requires a significant increase in area, which again would not be considered feasible from a financial point of view. Hence, the practical considered optimum areas for different loadings are shown in Table 9. It should be noted that these areas would give an effluent hydraulic loading of 292 mm per year, based on an effluent production of 100 L/c.day.

Table 8. Matrix showing the effect of varying the area on the mean number of overflow days simulated per year generated by four different dwelling sizes.

Area (m^2)	5PE	6PE	8PE	10PE
500	34	54	76	101
550	31	44	65	93
600	29	34	63	86
650	32	32	59	68
700	30	32	43	65
750	24	28	41	63
800	25	31	34	63
850	23	29	32	51
900	22	24	31	44
950	22	25	30	40
1000	17	23	28	34
1050	16	23	32	33
1100	15	17	31	31
1150	13	21	26	31
1200	12	17	24	29
1250	8	16	26	28
1300	7	15	24	32
1350	-	-	-	31
1400	-	-	-	30

Table 9. Practical optimum ET system area for different hydraulic loadings from a dwelling.

	Hydraulic loading					
	2PE	4PE	5PE	6PE	8PE	10PE
Optimum Area (m^2)	250	500	650	750	1000	1200

4.2.2. Free Surface Depth to Overflow

The optimum depth of free space above the surface of the soil maintained by a raised bund provides a significant amount of potential hydraulic storage although, ideally, long-term ponding of water on top of the willow system is not desirable. However, the organic and nutrient analysis carried out (see Section 3.4) showed that pollutant concentrations in any surface water were relatively low. Furthermore, no foul odours have been detected on any visits to the sites. The effect of varying the depth of free space on a 500 m^2 plan area willow system was conducted by entering a range of depths between 0 and 350 mm (50 mm increments) for three different overall willow system depths (1.5, 1.8, and 2.1 m). Overall depths greater than 2.1 m were not considered because of practical constraints during the excavation phase of construction. The change in the number of overflow days by varying the depth of free space is shown in Table 10.

Table 10. Effect of varying the depth of free space above the surface of the willow system on the mean number of overflow days simulated per year for three different system depths.

Depth of Free Space (mm)	Depth of ET System		
	1.5 m	1.8 m	2.1 m
0	76	65	59
50	72	61	55
100	62	47	40
150	58	44	38
200	42	35	34
250	37	33	34
300	33	32	28
350	30	29	27

Similar to the effect of increasing the area to lower the number of days with overflow, the gains made by increasing the free space appear to be less significant beyond a certain point. The optimum depth of free space in systems with a total depth of 2.1 and 1.8 m is approximately 200 mm (with 34 and 35 overflow days per year respectively), while the optimum depth for a system of depth 1.5 m is 300 mm (33 overflow days).

4.2.3. Hydraulic Load

The benefit of decreasing the hydraulic loading on the system by using an impermeable membrane on the surface of the willow system to direct a proportion of the rainfall away from the basin was also assessed. This analysis assumed that under such a scenario there would be no free storage space available above the soil. As the proportion of rainfall that would be diverted away from the willow system by using an impermeable membrane at the surface is unknown, a hypothetical analysis has been carried out where the rainfall is assumed to be reduced by varying degrees. The model again was based upon a willow system area of 500 m^2 receiving effluent from a dwelling with 4 PE with the results of the analysis shown in Table 11. This shows that the ET system could be modified to act as a fully zero-discharge treatment system if 60% or more of the annual rainfall falling onto the system is diverted. However, it should be recognised that the impact of covering the surface of the soil with an impermeable membrane may produce deleterious effects with respect to oxygen transfer down into the soil. This would also reduce soil evaporation.

Table 11. Matrix showing the mean number of overflow days simulated per year from the willow ET system for different reduced rainfall and depth combinations over a five-year period.

Rainfall	Depth of ET System						
	1.5 m	1.6 m	1.7 m	1.8 m	1.9 m	2.0 m	2.1 m
100%	76	73	68	65	64	62	59
90%	68	64	63	61	58	51	46
80%	58	55	51	43	39	36	34
70%	44	37	33	30	28	25	21
60%	28	24	19	16	13	12	9
50%	16	13	9	8	6	5	4
40%	6	4	3	1	0	0	0
30%	0	0	0	0	0	0	0
20%	0	0	0	0	0	0	0

4.2.4. Optimum Design (All Parameters)

In order to further refine the design parameters for an ET willow system, different combinations of plan area, overall depth, depth of gravel layer and depth of free space to overflow were varied to minimise the number of overflow days using the Excel Solver optimisation routine whereby the target parameter to be minimised was the number of overflow days. The constraints for the design parameters were as follows:

- Area (up to a maximum of the value detailed for the corresponding PE in Table 8).
- Depth (up to a maximum overall depth of 2.1 m due to practical constraints)
- Depth of gravel layer (up to a maximum of 350 mm due to cost constraints)
- Depth of free space at the surface (up to a maximum of 300 mm)

The optimum target parameter values (i.e., the smallest values) for different PEs are shown in Table 12. As would be expected, the area required to give the lowest number of overflow days is approximately equal to the maximum area constraint for all cases. The overall optimum depth is also equal to the maximum that the constraint in the model would allow (2.1 m) for all loading rates. There is some variance in the depths of the gravel layer determined for each the different PE, ranging from 279 mm to the maximum of 300 mm. The optimal free space depth above the surface is close to the maximum of 300 mm in all cases.

Table 12. Design parameters for willow system determined from optimisation.

PE	Area (m^2)	Overall Depth (mm)	Depth of Gravel Layer (mm)	Depth of Free Space (mm)	Overflow Days/yr
2	250	2095	300	295	30
4	498	2100	285	297	30
5	646	2083	279	298	30
6	750	2100	300	300	30
8	991	2100	295	295	30
10	1200	2100	300	300	31

4.2.5. Aspect Ratio

Regression analysis on the field trial results found no conclusive evidence that wind speed had a significant bearing on the ET rate from any willow system (ET_{willow}) with the exception of the 2011 ET_{willow} rate on Site B (p-value < 0.05) and the 2013 ET_{willow} rates on Site A being just outside the 95% confidence interval (p-value = 0.092). Equally, sensitivity analysis, in which the relative effect of the different meteorological parameters on ET_o was assessed, revealed that wind speed had a low Spearman Rank-order correlation coefficient with ET_o. Analysis also showed that wind speed had a relatively low influence on the ET_o: for example, a one standard deviation increase in wind speed would result in an increase of just 0.15 mm ET_o, compared to a one standard deviation increase in net radiation which would result in an increase of 1.38 mm ET_o. The effect of wind speed has been reported to have a greater effect on ET_o for taller crops although such research [20] has shown that the wind speed needs to be relatively high (>6 m/s) before any significant increase on expected crop evapotranspiration is gained. The average wind speed at the sites monitored as part of this research project was between 2 and 3 m/s. Hence, the field study results indicate that the impact of wind speed on ET_{willow} is not likely to be a critical factor in such a temperate, high humidity climate (such as Ireland's) and so any benefit derived from tailoring the willow system design to take advantage of wind speed via a large aspect ratio would be modest. However, if space is available on the site, then it is recommended to aim for a high aspect ratio design, as some benefit will be gained from the exposure to higher wind speeds throughout the willow standing.

5. DISCUSSION AND DESIGN CRITERIA

The optimal design dimensions for a willow ET system under different PE loadings (Table 11) indicate that some hydraulic overflow from the sealed systems would be expected during most years due to lower ET levels than the net hydraulic loading from effluent and rainfall. If the wastewater is introduced into the system via distribution pipes in the gravel layer at the base then the effluent will accumulate from the bottom up, compared to the rainwater on the surface which will accumulate from the top down. These hydraulic levels of effluent and rainfall in such an ET system have been simulated on a monthly basis across a hydrological year (i.e., the beginning of October to the end of September) to assess when an overflow would be expected and whether it would consist of primarily rainwater runoff or wastewater, as illustrated in Figure 10. This somewhat simple, yet instructive, analysis assumed that no mixing of effluent and rainwater occurred in the soil within the willow system but also that any water uptake by the willows is preferentially taken from the flooded level downwards (i.e., any ET will preferentially take stored monthly rainfall first before effluent). The depth of the ET system was set at 1.8 m with an overflow at 1.9 m in order to minimise the depth of any standing water and also encourage faster runoff of any rainfall events. The system was designed for a 4 PE residence producing effluent at a rate of 100 L/c.day. The analysis applied a water balance equation using mean annual monthly rainfall and other monitored meteorological data from the weather stations.

The results show that some overflow would be expected from the system during four months of the year (January–April). However, the respective levels of stored effluent from the base and rainfall from the surface indicate that any overflow should be rainwater runoff, which would thus present very little pollution threat to the environment. Indeed, the analysis shows that the majority of the water stored throughout the winter period would be rainwater. From May onwards, the wastewater would begin to make up the majority of stored water, but this would all be contained within the willow system with the maximum depth of wastewater only predicted to reach just over 1 m. It should be noted again that this analysis is based on the conservative assumption that the willow trees would preferentially evapotranspire the rainfall at the top before any deeper wastewater which is unlikely to be the case in reality, as the willow trees would have a preference for the nutrient rich wastewater.

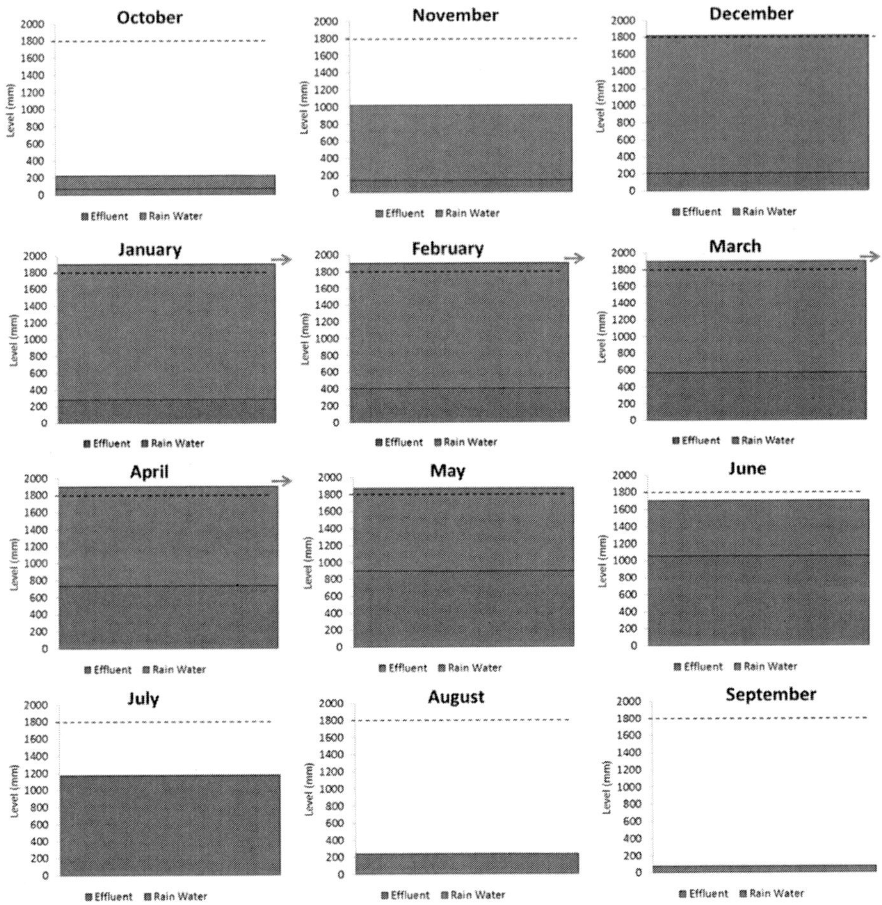

Figure 10. Expected water levels in willow ET evapotranspiration system across an annual cycle fed with effluent from the base.

Hence, this analysis provides a useful insight into the worst-case scenario that could occur and also demonstrates that the system can potentially operate effectively as a zero discharge systems with regards to the effluent loading. It is also important to recognise that this analysis was based upon an effluent production rate of 100 L/c.day (which equates to an aerial hydraulic loading of just less than 300 mm/year). This is similar to the effluent production measured on many of the sites throughout the monitoring period as well as in other research projects in Ireland [15,21] but significantly lower than the 150 L/c.day assumed in the national on-site wastewater Code of Practice [4] for a typical single dwelling. The effect of increasing the effluent loading rate to 150 L/c.day in the simulation is contained in Table 13 where it can be seen that the mean number of overflow days would increase by 60% to 80% for the same area of ET system for any given PE. If such a difference was to be negated by an increase in area, it would require an increase of between 40% and 50%, resulting in a very significant increase in cost.

Table 13. Design parameters for willow systems determined from optimisation.

PE	Area (m²)	Overflow Days per Year		Area	
		100 l/c.day	150 l/c.day	Increased Area (m²)	% increase
2	250	55	85	350	40%
4	500	55	85	675	35%
5	600	51	92	890	48%
6	750	55	85	1020	36%
8	1000	55	85	1350	35%
10	1200	51	92	1770	48%

Hence, it is imperative that, in such a climate where the expected annual ET_{willow} is going to be marginally higher than the expected annual rainfall at best, that wastewater production is kept as low as possible. Any design guidelines should include the requirement for modern water-efficient plumbing and appliances to be installed in the dwelling at the time of construction. It is also imperative that the rainfall runoff from the roof of the house is not plumbed into the wastewater system, as this would cause a significant increase in the hydraulic loading of the ET system. This practice appeared to be happening at Sites A and B with the corresponding enhanced hydraulic loads listed in Table 2.

6. CONCLUSIONS

A series of full-scale field trials on willow-based ET systems sited in areas of low permeability subsoil has produced data from which realistic ET values from such systems in the temperate, maritime, Irish climate have been gained. The field trials showed that none of the systems were able to evapotranspire all the effluent and rainfall hydraulic loading which were therefore deemed to be losing water by overflow or exfiltration at some times of the year. The effective porosities from the backfilled low porosity soils (i.e., storage space for effluent and rainfall) were also measured on site and found to be much lower than might typically be assumed for soil void ratios, ranging from just 9% to 20%. However, the quality of water stored in the systems, according to samples taken from the inspection wells, indicate that the systems act as very effective passive treatment systems. Furthermore, samples of ponded water on the surface of systems during winter periods have shown that the quality of any potential runoff would be similar to that of water in systems that had never received any effluent.

These field data have then been used to compare the influence of many design parameters on overall ET performance and so develop advice towards the formulation of generic national design guidelines. Realistic dimensions of ET systems for the treatment of wastewater from single houses have been established in order to minimize the numbers of days of any potential overflow. The plan areas of these systems are based on an annual effluent loading of just less than 300 mm per year. Further analysis has shown that if the effluent is fed from the base of the ET system designed to such a specification, this should ensure any overflow during winter periods will be predominantly rainfall runoff and that the net annual wastewater production is effectively fully evapotranspired.

ACKNOWLEDGMENTS

The research project was mainly financed by Wexford County Council, Ireland with some additional funding from the Environmental Protection Agency, Ireland. The authors also wish to acknowledge the contribution of Brendan Cooney (Wexford County Council) and Arne Backlund (Backlund ApS) to the research.

REFERENCES

1. Central Statistics Office (CSO). Census 2011, Principal Demographic Results; Central Statistics Office, Government of Ireland, Stationery Office: Dublin, Ireland, 2012.
2. Hynds, P.D.; Misstear, B.D.R.; Gill, L.W. Development of a microbial contamination susceptibility model for private domestic groundwater sources. Water Resour. Res. **2012**, 48.
3. Environmental Protection Agency (EPA). A Risk-Based Methodology to Assist in the Regulation of Domestic Waste Water Treatment Systems; EPA: Wexford, Ireland, 2013.
4. Environmental Protection Agency (EPA). Code of Practice: Wastewater Treatment and Disposal Systems Serving Single Houses; EPA: Wexford, Ireland, 2009.
5. Gregersen, P.; Brix, H. Zero-discharge of nutrients and water in a willow dominated constructed wetland. Water Sci. Technol. **2001**, 44, 407–412.

6. Arias, C.A. Current state of decentralized waste water treatment technology in Denmark. In Proceedings of the EPA International Symposium on Domestic Wastewater Treatment and Disposal Systems, Dublin, Ireland, 10–11 September 2012.
7. Miljøstyrelsen. Retningslinier for Etablering af Pileanlæg op til 30 PE. In Økologisk byfornyelse og spildevandsrensning, Nr. 25; Miljøstyrelsen: Copenhagen, Denmark, 2003.
8. Rosenqvist, H.; Aronsson, P.; Hasselgren, K.; Perttu, K. Economics of using municipal wastewater irrigation of willow coppice crops. Biomass Bioenergy **1997**, 12, 1–8.
9. Pauliukonis, N.; Shneider, R. Temporal patterns in evapotranspiration from lysimeter with three common wetland plant species in the eastern United States. Aquat. Bot. **2001**, 71, 35–46.
10. Elowson, S. Willow as a vegetation filter for cleaning of polluted drainage water from agricultural land. Biomass Bioenergy **1999**, 16, 281–290.
11. Dimitriou, I.; Aronsson, P. Wastewater and sewage sludge application to willows and poplars grown in lysimeters–Plant response and treatment efficiency. Biomass Bioenergy **2011**, 35, 161–170.
12. Kuzovkina, Y.; Knee, M.; Quigley, M. Soil compaction and flooding effects on the growth of twelve Salix L. species. J. Environ. Hort. **2004**, 22, 155–160.
13. Bialowiec, A.; Wojnowska-Baryla, I.; Agopsowicz, M. The efficiency of evapotranspiration of landfill leachate in the soil-plant system with willow Salix amygadlina. Ecol. Eng. **2007**, 30, 356–361.
14. Kele, B.; Midmore, D.J.; Harrower, K.; McKenniary, B.J.; Hood, B. An overview of the Central Queensland University self-contained evapotranspiration beds. Water Sci. Technol. **2005**, 51, 273–281.
15. Gill, L.W.; O'Luanaigh, N.; Johnston, P.M.; Misstear, B.D.R.; O'Suilleabhain, C. Nutrient loading on subsoils from on-site wastewater effluent, comparing septic tank and secondary treatment systems. Water Res. **2009**, 43, 2739–2749.
16. Department for Environment, Food and Rural Affairs (DEFRA). Growing Short Rotation Coppice, Best Practice Guidelines; Defra Publications: London, UK, 2002.
17. Caslin, B.; Finnan, J.; McCracken, A. Short Rotation Coppice Willow Best Practice Guidelines; Teagasc, Crops Research Centre: Carlow, Ireland, 2010.

18. Curneen, S.; Gill, L.W. Evapotranspiration Systems using willows for the treatment of on-site wastewater effluent in areas of low permeability subsoils. Ecol. Eng. **2015**. in press.

19. Guidi, W.; Piccioni, E.; Bonari, E. Evapotranspiration and crop coefficient of poplar and willow short-rotation coppice used as vegetation filter. Bioresour. Technol. **2008**, 99, 4832–4840.

20. Allen, R.G.; Pereira, L.S.; Raes, D.; Smith, M. Crop Evapotranspiration—Guidelines for computing crop water requirements. FAO Irrig. Drain. Paper **1998**, 55, 227.

21. Dubber, D.; Gill, L.W. Application of on-site wastewater treatment in Ireland and perspectives on its sustainability. Sustainability **2014**, 6, 1623–1642.

Chapter 9

Monitoring of Evapotranspiration in a Semi-Arid Inland River Basin by Combining Microwave and Optical Remote Sensing Observations

Guangcheng Hu [1] and Li Jia [1,2]

[1]State Key Laboratory of Remote Sensing Science, Institute of Remote Sensing and Digital Earth, Chinese Academy of Sciences, Beijing 100101, China

[2]Joint Center for Global Change Studies (JCGCS), Beijing 100875, China

ABSTRACT

As a typical inland river basin, Heihe River basin has been experiencing severe water resource competition between different land cover types, especially in the middle stream and downstream areas. Terrestrial actual evapotranspiration (ETa), including evaporation from soil and water surfaces, evaporation of rainfall interception, transpiration of vegetation canopy and sublimation of snow and glaciers, is an important component of the water cycle in the Heihe River basin. We developed a hybrid remotely sensed ETa estimation model named ETMonitor to estimate the daily actual evapotranspiration of the Heihe River basin for the years 2009–2011 at a spatial resolution of 1 km. The model was forced by a variety of biophysical parameters derived from microwave and optical remote sensing observations. The estimated ETa was evaluated using eddy covariance (EC) flux observations at local scale and compared with the annual precipitation and the MODIS ETa product (MOD16) at regional scale. The spatial distribution and the seasonal variation of the estimated ETa were analyzed. The results indicate that the estimated ETa shows reasonable spatial and temporal patterns with respect to the diverse cold and arid landscapes in the upstream, middle stream and downstream regions, and is useful for various applications to improve the rational allocation of water resources in the Heihe River basin.

Keywords: evapotranspiration; validation; spatial and temporal variations; ETMonitor; Heihe River basin

1. INTRODUCTION

The Heihe River basin is the second largest inland river basin in northwest China, which is characterized by its distinct cold and arid landscapes. Water consumption has increased due to the increase of population and cropland in the middle reaches, resulting in a decrease of incoming water and a series of environmental problems in the lower reaches [1,2]. To allocate water to different water use units based on water requirements and equity, the Heihe River Basin Irrigation water resource management Decision Support System (HD) has been developed and validated in the Zhangye city of the Heihe River basin [3]. Terrestrial actual evapotranspiration (ETa), including evaporation from soil and water surfaces, evaporation of rainfall interception, transpiration of vegetation canopy and sublimation of snow and glaciers, is an important component of the water cycle in the Heihe River basin. Accurate estimates of ETa is needed to reduce uncertainties in constructing weekly to monthly water balances across catchments and regions, which require knowledge of distribution of ETa at daily scales [4]. To better understand the eco-hydrological processes across the Heihe River basin, in particular the water resource and water use in the upstream, middle stream and downstream regions and their interactions, it is essential to characterize accurately the spatial and temporal patterns of land surface evapotranspiration in a continuous and quasi-operational manner.

Among algorithms estimating terrestrial actual evapotranspiration at regional scale, those using land surface biophysical parameters derived from satellite remote sensing observations are very promising due to their capability of spatial coverage and yet frequent repeat in time. Widely used algorithms can be divided into four categories: simplified empirical regression methods [5,6], trapezoid or triangle feature space methods [7,8,9], surface energy balance based (single- and dual-source) models [10,11,12,13,14], and the traditional evapotranspiration estimation approaches, for instance the Penman–Monteith equation and Priestley–Taylor equation, combined with remote sensing [4,15,16,17,18,19]. Efforts have been made using different models mainly based on energy balance equation and taking the evapotranspiration as the residual on way or the other in combination with optical remote sensing, i.e., using land surface temperature (LST) retrieved from thermal infrared bands as the main driving variable [20,21,22], of which some case studies were conducted in the Heihe River basin [9,23,24]. The main problem of the algorithms using LST is that the estimation approaches would be restricted to clear sky conditions. Besides, the physical meaning of the satellite-derived LST is unclear due to the mixed pixel problem, especially over heterogeneous and non-isothermal surfaces, and cannot be directly used as the aerodynamic temperature in estimating surface fluxes [25]. The estimated ETa as a residual of the energy balance contains biases from the net radiation, soil heat flux and sensible heat flux as well. Recently, some developments have been made based on the Soil–Vegetation–Atmosphere Transfer (SVAT) scheme, Penman–Monteith equation and remotely sensed data for the temporally and spatially continuous estimates of ETa, which is particularly useful for various applications since it is not limited to clear sky conditions [15,16,17,18,19,26].

The Penman–Monteith combination equation [27], which calculates evapotranspiration by combining the surface energy balance equation with other resistance-based methods, since it was proposed, has become a central component of very detailed evaporation equation incorporated in many land surface and atmospheric models [28,29]. The Shuttleworth–Wallace dual-source model [30] was proposed based on the Penman–Monteith equation for calculating soil evaporation and vegetation transpiration separately. The Shuttleworth–Wallace model is a biophysically sound framework for estimating terrestrial ETa at regional scale if the aerodynamic and surface resistances are well parameterized based on the remote sensing data and near-surface meteorological forcing data.

Considering the diverse landscapes and multi-climatic features in the Heihe River basin, and the demand of understanding the evapotranspiration processes for the whole river basin in different seasons, one should pay attention to the following issues: (1) The model should be robust enough to cover the processes of evaporation from bare soil and water surfaces, evaporation of rainfall interception, transpiration from grasses, crops and forests (perhaps also short vegetation understories beneath the forest canopy), sublimation of snow and glaciers. (2) Remote sensing data must be continuously available without gaps due to, for instance, clouds, instrument problems or other reasons. (3) Daily values of evapotranspiration are usually expected. In this study, we developed a hybrid remotely sensed ETa estimation model named ETMonitor to estimate the daily actual evapotranspiration of the Heihe River basin for the years 2009–2011 at a spatial resolution of 1 km. The estimation of soil evaporation and vegetation transpiration was based on the Shuttleworth–Wallace model with a variety of parameterization methods of the aerodynamic and bulk surface resistances, and the rainfall interception was estimated using a reformulated remote sensing based Gash model. The ETMonitor model uses a variety of biophysical parameters (except for LST) derived from microwave and optical remote sensing observations as input data to estimate the daily ETa for all sky conditions, which is particularly useful for routine applications in the Heihe River basin. The estimated ETa was evaluated using the in-situ observations and MODIS evapotranspiration product MOD16 at 3 eddy covariance (EC) flux sites located in the upstream and middle stream regions, and the spatial and temporal characteristics of ETa in the Heihe River basin were analyzed.

2. Study Area

The Heihe River basin is an inland river basin located in the arid region of northwest China, covering an area of approximately 1.432×10^5 km^2 (Figure 1). The landscape of the basin is characterized by very different land cover types and different climate zones in upstream, middle stream and downstream regions:

1. The upstream of the basin is a semi-humid, cold and high elevated mountainous region with the terrain elevation ranging from 2000 to 5000 m and the highest elevation of 5564 m above the sea level. The dominant land cover types are mountain forest (Picea crassifolia, Sabina przewalskii), alpine meadow, frozen soil, snow and glaciers. The annual precipitation is more than 350 mm/year and the mean annual air temperature is 1.5 °C–2.0 °C.

2. The middle stream area of the Heihe River basin, with an elevation of 1300–1700 m, is characterized by typical irrigated artificial oasis and desert with semi-arid climate. The primary crops are maize, spring wheat, vegetables and small fraction of orchards. The annual precipitation in the middle reaches is 100–250 mm/yr and the mean annual air temperature is 2.8 °C–7.6 °C.

3. The downstream area of the basin is well known as arid Gobi desert with an elevation of 900–1300 m. The vegetation cover mainly distributed as narrow strips (river greenbelt) along the river channel and a small oasis (named Ejin Banner Oasis) at the tail of the river. The dominant types of vegetation in the downstream region are Populus euphratica, Tamarix chinensis, and agricultural crops in the Ejin Oasis and along the river greenbelt. A terminal lake named East Juyanhai is located at the end of the river surrounded by desert. The annual precipitation in the downstream area is less than 50 mm/yr, and the mean annual air temperature is about 8.0 °C.

The water resource of the Heihe River is mainly from the rainfall, snow and glaciers in the upstream mountainous areas. Irrigation water used in the middle reaches agricultural area is mainly from the Heihe River through irrigation canals (main canal, branch canal, lateral canal, and field canal). Groundwater depletion is increasing year by year due to insufficient water supply from the river. Intensive use of water for crop irrigation in growing season over the middle reaches results in less or even cut-off of river water flow to the lower reaches every year. This, on one hand, influences the vegetation development in the downstream area, and on the other hand also perturbs the recharge of groundwater in the lower reaches. The ecosystem in the downstream area of the Heihe River basin is therefore under vulnerable conditions.

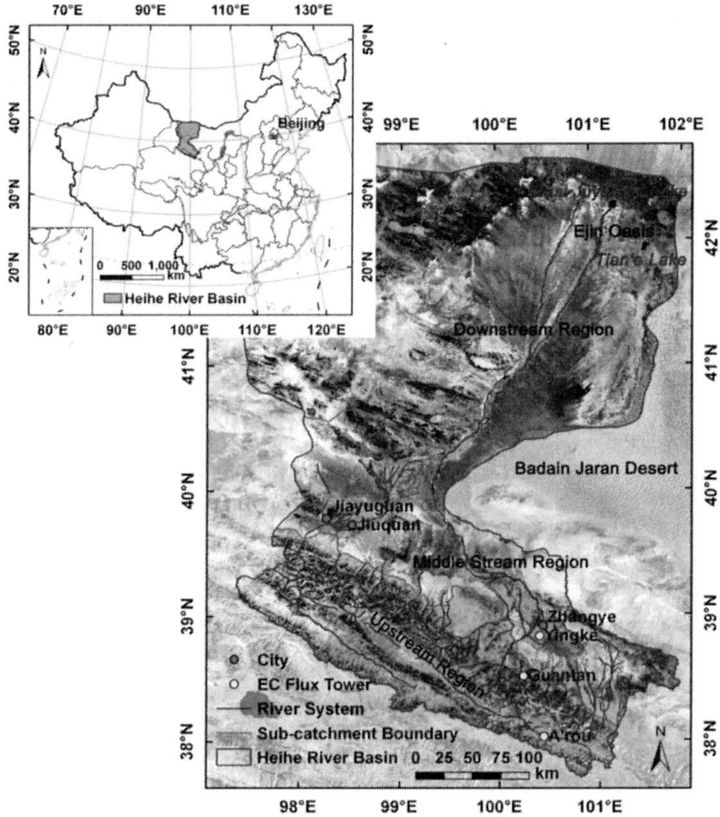

Figure 1. Locations of the Heihe River basin and the eddy covariance (EC) flux sites. The color composite remote sensing imagery of the landscape is mosaicked from Landsat-5 Thematic Mapper (TM) and Landsat-7 Enhanced Thematic Mapper Plus (ETM+) sensors (Red/band 7, Green/band 4, and Blue/band 2).

3. THEORETICAL FORMULATION OF THE ETMONITOR MODEL

The ETMonitor model combines different ETa parameterizations for the following land cover types (Figure 2): (1) water body; (2) snow/ice surface; and (3) soil–vegetation canopy. For different vegetation types, the evaporation of rainfall interception is also taken into account in ETMonitor. The theoretical formulation of the ETMonitor will be presented in this section, and the forcing data will be briefly introduced in Section 4.

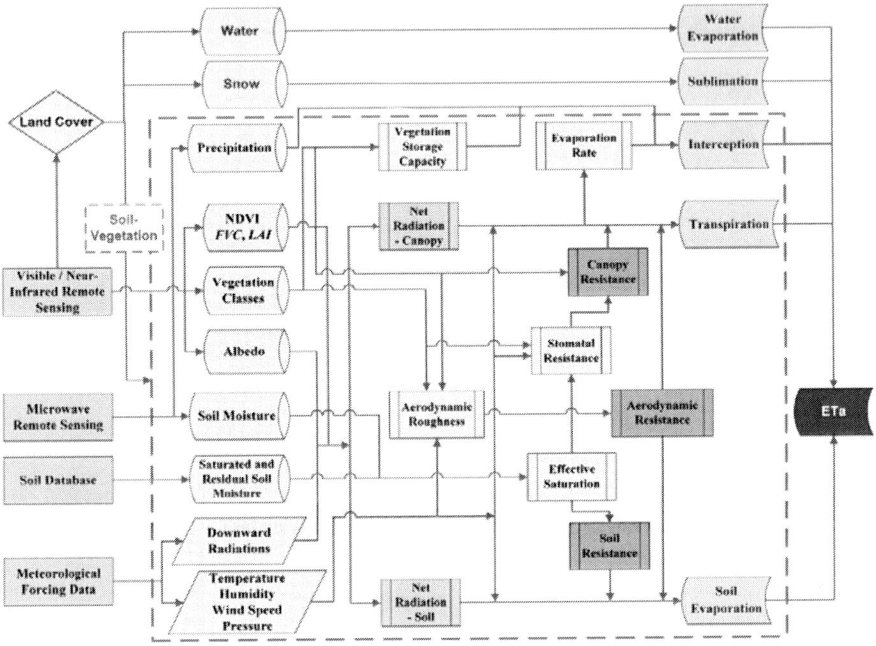

Figure 2. Flowchart of the ETMonitor model combining different evapotranspiration (ETa) parameterizations.

3.1. Evapotranspiration of the Soil–Vegetation Unity

3.1.1. Shuttleworth–Wallace Dual-Source Model

For an incomplete (partial) soil–vegetation canopy, the model is constructed with a dual-source scheme that distinguishes the energy partitioning and water fluxes between soil and vegetation (i.e., evaporation from the soil and transpiration from the vegetation canopy). The model structure together with its resistance scheme is illustrated in Figure 3.

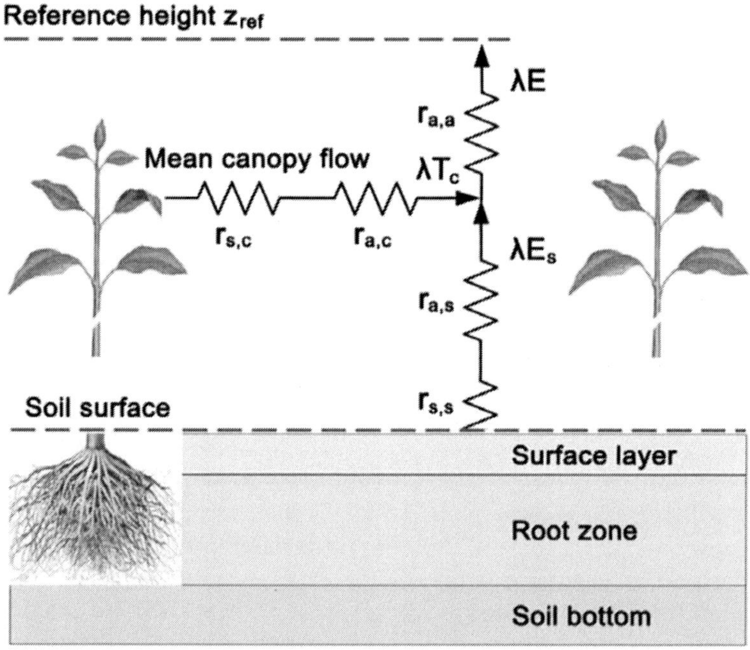

Figure 3. Schematic description of the latent heat flux partitioning with the relevant resistances for partial soil–vegetation canopy used in ETMonitor.

The Shuttleworth–Wallace dual-source model [30] is adopted to account separately for the energy balance for vegetation and soil components of a soil–vegetation canopy unity. Assuming aerodynamic mixing occurring at a mean canopy source height within the canopy, a variant form of Penman–Monteith equation that combines energy balance with plant physiology and soil water diffusion can be written for λE:

$$\lambda E = C_c \cdot PM_c + C_s \cdot PM_s \tag{1}$$

$$PM_c = \frac{\Delta(R_n - G) + [\rho C_p VPD - \Delta r_{a,c}(R_{n,s} - G)]/(r_{a,a} + r_{a,c})}{\Delta + \gamma[1 + r_{s,c}/(r_{a,a} + r_{a,c})]} \tag{2}$$

$$PM_s = \frac{\Delta(R_n - G) + (\rho C_p VPD - \Delta r_{a,s} R_{n,c})/(r_{a,a} + r_{a,s})}{\Delta + \gamma[1 + r_{s,s}/(r_{a,a} + r_{a,s})]} \tag{3}$$

$$C_c = \{1 + R_c \cdot R_a / [R_s \cdot (R_c + R_a)]\}^{-1} \tag{4}$$

$$C_s = \{1 + R_s \cdot R_a / [R_c \cdot (R_s + R_a)]\}^{-1} \tag{5}$$

$$R_a = (\Delta + \gamma) \cdot r_{a,a} \tag{6}$$

$$R_s = (\Delta + \gamma) \cdot r_{a,s} + \gamma \cdot r_{s,s} \tag{7}$$

$$R_c = (\Delta + \gamma) \cdot r_{a,c} + \gamma \cdot r_{s,c} \tag{8}$$

where λE ($W \cdot m^{-2}$) is the bulk latent heat flux of the soil–vegetation unity and λ ($J \cdot kg^{-1}$) is the latent heat of evaporation. PM_c and PM_s ($W \cdot m^{-2}$) are variables related to canopy transpiration and soil evaporation, C_c and C_s (dimensionless) are the surface resistance coefficients for canopy and soil, respectively. G ($W \cdot m^{-2}$) is the conductive heat flux from soil surface to deeper soil layer, R_n ($W \cdot m^{-2}$) is the total net radiation flux, $R_{n,c}$ and $R_{n,s}$ ($W \cdot m^{-2}$) are net radiation flux absorbed by the vegetation canopy and arrives at the soil surface, respectively, which can be estimated based on the remotely sensed albedo and fractional vegetation cover (FVC). $\Delta = d(e_{sat})/dT \cdot (Pa \cdot K^{-1})$, which is the slope of the curve relating saturated water vapor pressure e_{sat} (Pa) to temperature T (K), ρ ($kg \cdot m^{-3}$) is the air density, C_p ($J \cdot kg^{-1} \cdot K^{-1}$) is the specific heat of air, $\gamma \cdot (Pa \cdot K^{-1})$ is the psychrometric constant, VPD (Pa) is the vapor pressure deficit of the air at the reference height above the canopy. The terms R_a, R_s, and R_c (Pa $K^{-1} \cdot s \cdot m^{-1}$) are the total aerodynamic resistances to vapor transport for the whole soil–vegetation canopy, the soil and the vegetation, respectively, that are derived from the surface resistance and aerodynamic resistance. $r_{a,c}$ ($s \cdot m^{-1}$) is the boundary layer aerodynamic resistance between the canopy and the canopy source height, $r_{s,c}$ ($s \cdot m^{-1}$) is the bulk canopy surface resistance, $r_{a,s}$ ($s \cdot m^{-1}$) is the aerodynamic resistance between the soil surface and the canopy source height, $r_{s,s}$ ($s \cdot m^{-1}$) is the soil surface resistance. The $r_{a,a}$ ($s \cdot m^{-1}$) illustrated in Figure 3 is the aerodynamic resistance between the canopy source height and the reference height above the canopy. The canopy source height is taken at d + z_{0m} (m), where d (m) and z_{0m} (m) are zero plane displacement height and aerodynamic roughness length for momentum transfer of the soil–vegetation canopy, respectively.

3.1.2. Resistances

There have been numerous studies and achievements on parameterization methods of the above resistances since the 1970s based on field experiments and modeling at field scale over different vegetation and soil management regimes. ETMonitor adopts resistance equations and corresponding parameters from previous studies for different landscapes, which are described below together with the relevant literatures.

(1) Aerodynamic Resistance

Three aerodynamic resistances, i.e., $r_{a,a}$, $r_{a,c}$, and $r_{a,s}$ ($s \cdot m^{-1}$), used in the ETMonitor model (as shown in Figure 3) are taken from [31]:

$$r_{a,a} = \frac{u_{zref} - u_c}{u_{zref}} r_a \tag{9}$$

$$r_{a,c} = \frac{u_c}{\sigma_a u_{zref}} r_a \tag{10}$$

$$r_{a,s} = \frac{u_c}{(1-\sigma_a) u_{zref}} r_a \tag{11}$$

where u_{zref} ($m \cdot s^{-1}$) is the wind speed at the reference height z_{ref} (m) above the canopy, u_c ($m \cdot s^{-1}$) is the wind speed at the canopy source height ($d + z_{0m}$) (m), r_a ($s \cdot m^{-1}$) is the overall aerodynamic resistance for momentum transfer in the soil–vegetation unity:

$$r_a = \frac{1}{k^2 u_{zref}} [\ln(\frac{z_{ref} - d}{z_{0m}}) - \Psi_m][\ln(\frac{z_{ref} - d}{z_{0h}}) - \Psi_h] \tag{12}$$

where k is von Karman coefficient (0.41), ψ_m and ψ_h are stability correction functions (neutral condition is assumed at daily step of calculation so as $\psi_m = 0$ and $\psi_h = 0$). The roughness length for heat transfer z_{0h} (m) is set as $0.1z_{0m}$ (m).

In Equations (10) and (11) σ_a is a momentum partition coefficient (i.e., momentum absorption is assumed to take place partly by the canopy and partly by the soil) assumed to be a function of Leaf Area Index (LAI) [32] and given by:

$$\sigma_a = 1 - \frac{0.5}{0.5 + LAI} \exp(-\frac{LAI^2}{8}) \tag{13}$$

The u_c ($m \cdot s^{-1}$) in Equations (9)–(11) is parameterized as:

$$u_c = 0.83 \sigma_f u_{zref} + (1 - \sigma_f) u_{zref} \tag{14}$$

where σ_f is an extinction coefficient, ranging between 0 (for bare soil) and 1 (for full vegetation cover) [33]. In ETMonitor, σ_f is taken as the same value as σ_a, which depends on LAI and covers the same range of values, satisfying the gradual transition from bare soil to a closed canopy [31].

(2) Soil Surface Resistance

Following Camillo et al., Clapp et al. and Dolman [34,35,36], the soil surface resistance to vapor flux $r_{s,s}$ ($s \cdot m^{-1}$) is parameterized as a function of minimum soil surface resistance (under optimal water condition) $r_{s,s,min}$ ($s \cdot m^{-1}$) and relative water content of soil surface layer θ_g' (dimensionless):

$$r_{s,s} = r_{s,s,min} (\theta_g')^{-3} \tag{15}$$

where $r_{s,s,min}$ is taken as 50 $s \cdot m^{-1}$. The relative water content of soil surface layer θ_g' is calculated as a degree of saturation of the surface layer volumetric soil water content θ_g ($cm^3 \cdot cm^{-3}$) [37] derived from the observations by the passive microwave radiometer, i.e., the Advanced Microwave Scanning Radiometer–Earth Observing System (AMSR-E):

$$\theta_g' = \frac{\theta_g - \theta_{res}}{\theta_{sat} - \theta_{res}} \tag{16}$$

where θ_{sat} and θ_{res} ($cm^3 \cdot cm^{-3}$) are saturated and residual soil water contents which are soil types dependent.

(3) Canopy Surface Resistance

The canopy surface resistance is estimated by scaling up leaf stomatal resistance with canopy leaf area. The stomatal resistance r_{st} ($s \cdot m^{-1}$) depends on ambient factors (available energy and atmospheric water demand) and available soil moisture in the root zone. Taking $r_{st,min}$ ($s \cdot m^{-1}$) as the minimum stomatal resistance (defined as the value of stomatal resistance under fully open stomata conditions with sufficient energy and water supply), the canopy surface resistance $r_{s,c}$ ($s \cdot m^{-1}$) is parameterized by a multiplicative response to several environmental factors [38,39,40],

$$r_{s,c} = \frac{F_s}{LAI} \frac{r_{st,min}}{f_1(R_s) f_2(T_a) f_3(VPD) f_4(\theta_r)} \tag{17}$$

where LAI ($m^2 \cdot m^{-2}$) is the leaf area index, F_s is the leaf shadowing factor and a function of LAI and canopy structure, $f_1(R_s)$, $f_2(T_a)$, $f_3(VPD)$ and $f_4(\theta_r)$ are functions (limiting factors) that describes stomatal response to environmental factors of solar radiation R_s ($W \cdot m^{-2}$), air temperature T_a (°C), vapour pressure deficit VPD (hPa) and root zone soil water content θ_r ($cm^3 \cdot cm^{-3}$), respectively.

Following Mehrez et al. [41] the leaf shadowing factor F_s is linked to LAI by:

$$F_s = 0.3LAI + 1.2 \tag{18}$$

The limiting factors f_i(variables) (i = 1, 2, 3, 4) are adopted from literatures [40,42,43,44,45] and given below:

$$f_1(R_s) = 1 - \exp(-\frac{R_s}{k_{Rs}}) \tag{19}$$

$$f_2(T_a) = \frac{(T_a - T_L)(T_H - T_a)^a}{(T_{opt} - T_L)(T_H - T_{opt})^a} \tag{20}$$

$$a = \frac{T_H - T_{opt}}{T_{opt} - T_L} \tag{21}$$

$$f_3(VPD) = 1 - k_{VPD} VPD \tag{22}$$

where k_{Rs} is taken as 500 W·m^{-2}, k_{VPD}, T_L, T_{opt} and T_H depends on vegetation types. For k_{VPD}, Noilhan et al. [46] has derived a value of 0.025 hPa^{-1} for a coniferous forest, Stewart [40] has obtained a value of 0.0155 hPa^{-1} for rainfed grass prairie, and a value of 0.023 hPa^{-1} is taken for crops in ETMonitor. T_L, T_{opt} and T_H are the lower, optimum and upper air temperature limits of stomatal activity, which can be taken as 5 °C, 35 °C and 55 °C for maize (C_4 carbon fixation), 0 °C, 25 °C and 50 °C for the other crop varieties and grasses (C_3 carbon fixation), 0 °C, 20 °C and 40 °C for the forests, respectively.

A sinusoidal function is used to take into account the influence of root zone soil moisture deficit on the stomatal closure [15,47],

$$f_4(\theta_r) = K_{sf}\theta_r^{'} - \frac{\sin(2\pi\theta_r^{'})}{2\pi} \tag{23}$$

$$\theta_r^{'} = 0.1LAI + (1 - 0.1LAI)\{1 - \exp[\theta_g^{'}(-0.5LAI - 1)]\} \tag{24}$$

where $\theta_r^{'}$ (dimensionless) is the relative water content of the root zone. The parameter K_{sf} is a tenacity factor that describes the ability of plants to extract soil moisture and ranges from 1 for drought sensitive plants to 1.5 for moderately drought sensitive plants and to 3 for insensitive plants.

The limiting factors f_i(variables) have values between 0 and 1 (0 for full closure of stomata which leads to infinite resistance; 1 for fully open stomata which leads to minimum stomatal resistance corresponding to optimum conditions). When all environmental conditions are under optimum, all the f_i(variables) are equal to unity and the stomatal resistance r_{st} takes the minimum value of $r_{st,min}$. Under conditions away from the optimum, the stomatal resistance is increased from its minimum value $r_{st,min}$. When one or more of the f_i(variables) reach zero, Equation (17) becomes infinite resistance value, in this case a maximum value of 5×10^4 $s \cdot m^{-1}$ is given to $r_{s,c}$.

3.2. Interception

Interception is the rainfall intercepted by the vegetation canopy and subsequently evaporated back to the atmosphere. We estimated the interception of different vegetation types in the Heihe River basin using a reformulated Gash analytical model [48,49] called RS–Gash model (remote sensing based Gash model). The reformulated Gash model is forced by remote sensing observations of canopy structure (e.g., FVC, LAI, vegetation storage capacity) and rainfall intensity [50,51]. The interception of canopy and trunk was replaced by the interception of vegetation, and the sub-pixel heterogeneity was taken into account by applying a Poisson distribution function to the LAI value of each pixel. The detailed model description and validation is given in parallel work by Cui and Jia [50] and Cui et al. [51].

3.3. Sublimation

Sublimation of snow and ice (glaciers) occurs more readily when the near-surface water vapor pressure is less than the saturated water vapor pressure. The sublimation of snow and ice was estimated using the formula developed by Kuzmin [52] and recommended for application by the World Meteorological Organization (WMO):

$$E_{ice} = (0.18 + 0.098u_{10})(e_s - e_2) \tag{25}$$

where E_{ice} ($mm \cdot d^{-1}$) is the sublimation of snow and ice, u_{10} ($m \cdot s^{-1}$) is the wind speed at a height of 10 m above the surface, e_2 (hPa) is the actual water vapor pressure at a height of 2 m, e_s (hPa) is the saturated water vapor pressure.

3.4. Evaporation from Water Surface

The classical Penman equation [53] was used to estimate the evaporation from water surface:

$$\lambda E_{\text{water}} = \frac{\Delta(R_n - G) + \gamma E_a}{\Delta + \gamma} \tag{26}$$

$$E_a = 6.43(a_w + b_w u_2)(e_s - e_2) \tag{27}$$

where λE_{water} is the latent heat of evaporation in $MJ \cdot m^{-2} \cdot day^{-1}$, λ is as defined in Equation (1) in $MJ \cdot kg^{-1}$, R_n and G are as defined in Equation (2) in $MJ \cdot m^{-2} \cdot day^{-1}$, Δ and γ are as defined in Equation (2) in $kPa \cdot K^{-1}$, e_2 and e_s are as defined in Equation (25) in kPa. u_2 ($m \cdot s^{-1}$) is the wind speed at a height of 2 m, a_w and b_w are wind function coefficients, which are taken as 0.5 and 0.54 for open water.

4. DATA

4.1. Remote Sensing Data

In the ETMonitor model, the biophysical variables derived from microwave and optical remote sensing observations include albedo, FVC, LAI, surface layer volumetric soil water content θ_g, and land cover types, which are described separately in this section.

4.1.1. Land Surface Albedo

Land surface albedo is a critical variable affecting the radiative energy budget and the convective overturning. The albedo data used in this study is the Global LAnd Surface Satellite (GLASS) albedo product from 2009 to 2011, which is produced at 1 km spatial resolution and in eight-day steps in the Sinusoidal projection [54]. The GLASS albedo product is a gapless and long-term continuous data set with an accuracy similar to that of the Moderate Resolution Imaging Spectroradiometer (MODIS) MCD43 product.

4.1.2. Normalized Difference Vegetation Index (NDVI)

The NDVI was used to indicate the vegetation conditions and to calculate the FVC [33] and LAI [15]. The NDVI data used in this study is MODIS vegetation index (VI) product MOD13A2 (1-km/16-day) from 2009 to 2011 in the Sinusoidal projection. Cloud presence usually contaminates remotely sensed data, especially for the optical and thermal infrared bands. MODIS NDVI data is provided at 16-day compositing periods utilizing three compositing schemes to reduce the artifacts due to clouds and aerosols. For some time periods that cloud cover can persist longer than the compositing period; these compositing schemes cannot eliminate atmospheric contamination. Therefore a time series reconstruction algorithm proposed by Jia et al. [55] based on the Harmonic Analysis of Time Series (HANTS) approach and the Temporal–Similarity–

Statistics (TSS) method was implemented to obtain daily cloud-free NDVI time series being consistent at spatial and temporal scales.

4.1.3. Soil Moisture

Soil moisture is a key variable in understanding land surface hydrology and controls the ecological, hydrometeorological and biogeochemical processes. Land surface soil moisture was derived from the passive microwave radiometer AMSR-E on the Aqua satellite. Daily surface soil moisture product at 25 km spatial resolution was obtained from the National Snow and Ice Data Center (NSIDC). The daily soil moisture for every four days were averaged to obtain four-day aggregated soil moisture and then downscaled to 1 km using the bilinear resampling method.

4.1.4. Land Cover

Land cover type is a crucial parameter that is required for creating look-up tables about certain biophysical properties, such as the minimum stomatal resistance and the optimal air temperature for the growth of the plant. The Multi-source Integrated Chinese Land Cover (MICLCover) data at 1 km spatial resolution used in this study was generated by combining multi-source Land Use/Land Cover (LULC) products based on the Dempster–Shafer evidence theory [56]. The land cover scheme identifies 17 land cover types defined by the International Geosphere–Biosphere Programme (IGBP). The crop varieties (maize, spring wheat, and potato) were also accounted for using the land cover map provided by the Heihe Plan Science Data Center [57]. The water, permanent snow and glaciers mask was derived from the MODIS land cover type product MCD12Q1 from 2009 to 2011 to indicate the yearly dynamics of surface water, snow and glaciers.

4.1.5. Snow Cover Extent

Snow cover plays a significant role in land surface models and hydrometeorology studies. In this study, the MODIS remotely sensed snow cover extent product was used as model input, i.e., the maximum snow cover extent over an eight-day compositing period at 500 m spatial resolution derived from the MODIS/Terra Snow Cover (MOD10A2) data set. The MODIS snow cover data is derived using a snow-mapping algorithm based on the Normalized Difference Snow Index (NDSI) and other criteria tests for snow reflectance characteristics in the visible and near-infrared regions.

4.1.6. Precipitation

The version 7 of Tropical Rainfall Measuring Mission (TRMM) Multi-satellite Precipitation Analysis (TMPA) product 3B42V7 was used to estimate the interception. The 3B42V7 with the spatial and temporal resolutions of 0.25°/tri-hourly over 50°N–50°S is available from 1998 to present. The daily precipitation was obtained by summing up all of the corresponding tri-hourly values for each day, and then spatially interpolated onto the MODIS grid using a bilinear interpolation algorithm.

4.2. Soil Texture and Hydraulic Parameters

The saturated and residual soil water contents of soil surface layer were provided by the China dataset of soil hydraulic parameters [58]. The dataset is derived from multi-PTFs (Pedotranfer Functions), including the parameters in the Clapp and Hornberger Functions (FCH) and in the van Genuchten and Mualem Functions (FGM), respectively. It is available at 30 arc-second (1 km) and seven vertical layers to the depth of 1.38 m.

4.3. Meteorological Forcing Data

The near-surface meteorological forcing data, including air temperature, air pressure, specific humidity, wind speed, downward short-wave and long-wave radiation fluxes are essential to drive the ETMonitor. The 5 km/1 h meteorological forcing data from 2009 to 2011 over the Heihe River basin used in this study was derived based on the Weather Research and Forecasting model (WRF) simulation [59,60].

The 5 km/1 h meteorological forcing data was statistically downscaled to the spatial–temporal resolution of 1 km/daily. First, the hourly data was averaged to daily for every meteorological parameter. Second, the spatial resolution of 5 km was downscaled to 1 km using statistical downscaling approaches for daily air temperature, air pressure, specific humidity, downward short-wave and long-wave radiations. The statistical downscaling is based on the semi-empirical relationships and on the multiple predictor variables (topographic altitude, aspect, and slope, at 1 km and 5 km spatial resolutions) and transfer functions [61,62]. Then the wind speed was downscaled to 1 km using the bilinear resampling method.

4.4. Eddy Covariance Flux Data

The estimated ETa by ETMonitor was evaluated using the ground measurements across different ecosystems at three eddy covariance flux sites in 2009–2011. The latter were collected from the Watershed Allied Telemetry Experimental Research (WATER) and provided by the Cold and Arid Regions Sciences Data Center at Lanzhou (available at http://westdc.westgis.ac.cn/ (accessed on 1 November 2014)) [63,64,65]. The ground measurements of latent heat flux with a time interval of 30-minute by the eddy covariance method at the towers were measured and processed. The 30-minute flux data were obtained from the raw data sampled at 10 Hz using the post-processing software EdiRe, and then the half-hourly results were screened following the criteria proposed by Blanken et al. [66]. The quality controlled 30-minute latent heat flux was then aggregated to daily and eight-day values for comparison with the estimated ETa at consistent temporal intervals.

The three EC flux sites include Yingke, A'rou and Guantan, representing different climate and land cover types in the Heihe River basin (Table 1). The Yingke site is located in the middle reaches of the basin, and the other two sites are located in the upper reaches (Figure 1). The source areas of the EC measurements were within a radius of 250 m at these three sites [65]. The Energy Balance Ratio (EBR), i.e., the ratio of the turbulent energy fluxes (the sum of sensible and latent heat fluxes) to the available energy (net radiation flux minus soil surface heat flux) was used to indicate the energy balance closure at these three EC sites [65], and the values were similar to that (approximately 70%–90%) previously observed for cropland, grassland and forest sites [67,68,69].

Table 1. Eddy covariance flux sites of the Watershed Allied Telemetry Experimental Research (WATER) used for ETa evaluation at local scale.

Site Name	Sensors	Location/Elevation	Landscape	Data DOI
Yingke	CSAT3 (Campbell) and Li7500 (Li-cor)	100°24′37.2″E 38°51′25.7″N 1519.1 m	Cropland (maize)	10.3972/water973.0278.db
A'rou		100°27′52.9″E 38°02′39.8″N 3032.8 m	Alpine Meadow	10.3972/water973.0282.db
Guantan		100°15′00.8″E 38°32′01.3″N 2835.2 m	Spruce Forest	10.3972/water973.0294.db

4.5. MOD16 ETa

We also compared the estimated ETa by ETMonitor with the MODIS global evapotranspiration product MOD16, which is the first published 1 km ETa data set for the global vegetated land areas at eight-day, monthly and annual intervals [18,19] partly driven by remote sensing data. In the MOD16 product, the calculation of ETa were not made over non-vegetated pixels, i.e., water body, barren or sparsely vegetated, permanent snow and ice, permanent wetland and urban areas, but assigned by filled values. The data used for the comparative study is the MOD16A2 ETa product, which is produced at 1 km spatial resolution and eight-day compositing periods (mm/8-day) in the Sinusoidal projection.

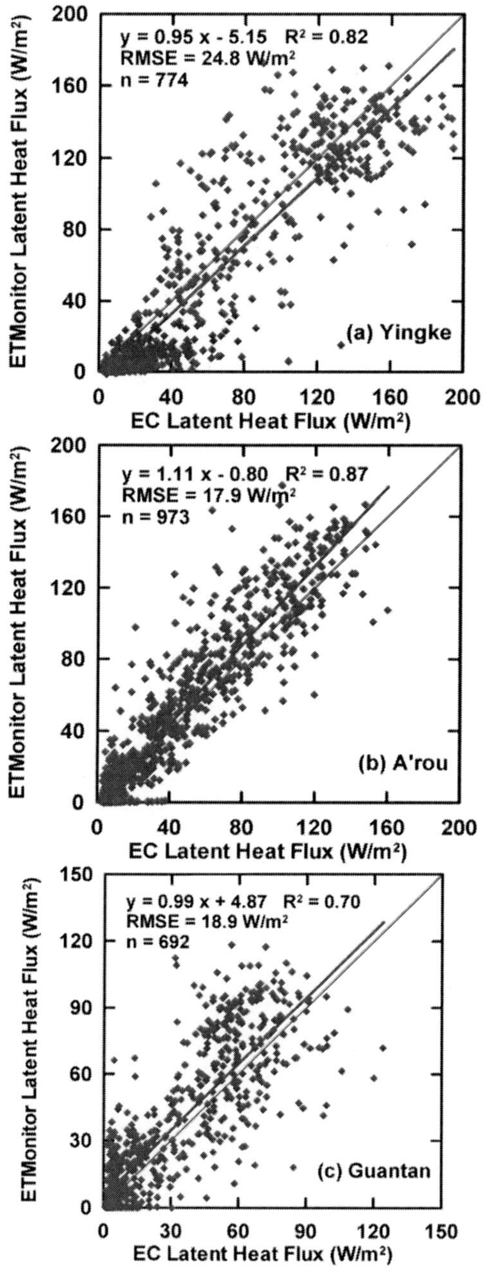

Figure 4. Scattering plots of the ETMonitor latent heat flux (λE) against eddy covariance (EC) observations spanning 2009–2011 at the site of (**a**) Yingke (cropland), (**b**) A'rou (alpine meadow), and (**c**) Guantan (forest). Values shown are the daily values of latent heat flux, λE (W/m^2).

5. RESULTS

5.1. Validation of the Model

5.1.1. Comparison with the Ground Measurements

Three-year (2009–2011) time series of daily latent heat flux λE (W/m^2, or ETa when expressed in mm/day) of the Heihe River basin was derived by the ETMonitor model. In order to evaluate the accuracy of the ETMonitor model, the estimated daily λE was compared with the EC observations at the site of Yingke (cropland), A'rou (alpine meadow), and Guantan (forest) separately (Figure 4). For such comparison, one should keep in mind that the estimated pixel values of λE might be different from in-situ measurements for the obvious spatial scale difference between the ground measurements and the satellite pixels [70,71,72]. The source areas of the EC measurements were within a radius of 250 m at these three sites [65]. ETMonitor performs the best at A'rou, then followed at Yingke with the coefficient of determination (R^2, the square of the correlation coefficient between ETMonitor and EC measurements) higher than 0.8. At Guantan, the discrepancy between estimated λE and EC observations is relatively large with the R^2 of 0.7. The root mean square error (RMSE) at the Yingke site is 24.8 W/m^2, a little higher than that of the other two sites.

We additionally compared the ETMonitor output with the MOD16 ETa at these three EC sites. The estimated daily ETa was aggregated to eight-day means and the time series plots and scatter plots at each site are shown in Figure 5 and Figure 6 to provide the assessment of the seasonal performance and statistical results of ETMonitor and MOD16 ETa. At the Guantan site, the missing EC data for most of 2009 was due to the deficiency of solar power, instrument malfunction, poor maintenance, and bad weather conditions (Figure 5c). In general, fairly good results have been obtained from this validation, showing that the ETMonitor model can reproduce the time series of ETa observed at local scale to some degree for these three landscapes. At Yingke in the spring season, the ETMonitor ETa was obviously underestimated due to the underestimated surface layer soil moisture (Figure 5a). The cropland in the oasis was irrigated in late autumn of the previous year, and then the water was stored in the frozen soil. In spring, the liquid soil water content is very high during thawing of frozen soil, enhancing the process of soil evaporation. However, such a dampened process was not captured by the AMSR-E soil moisture data used in this study. On one hand the AMSR-E soil moisture with rather coarse spatial resolution is a mixing contribution from different land surface elements, which may have smoothed out the spatial variability in land surface soil moisture. On the other hand, other studies have revealed that the AMSR-E soil moisture product exhibits underestimates [73,74]. The MOD16 ETa

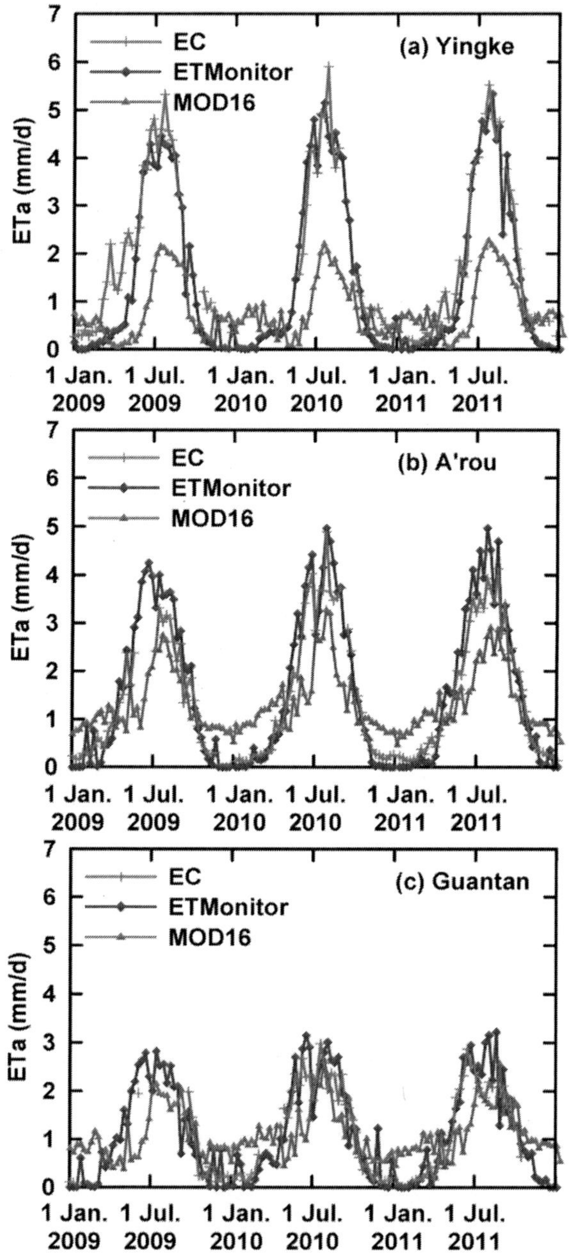

Figure 5. Three-year eight-day time series of ETMonitor, MOD16 and eddy covariance (EC) observed evapotranspiration (ETa) spanning 2009–2011 at the site of (**a**) Yingke (cropland), (**b**) A'rou (alpine meadow), and (**c**) Guantan (forest). Values shown are the eight-day means of ETa (mm/day).

is overestimated in winter, whereas in the growing season, it is much lower than EC observations at these three sites, especially at the Yingke Oasis station (Figure 5a). In general, the ETMonitor estimated ETa is closer to the EC observations than MOD16 with better correlation, smaller RMSE between ETMonitor and EC measurements. The R^2 of ETMonitor vs. EC (MOD16 vs. EC) is 0.92 (0.64), 0.96 (0.71), 0.87 (0.41) and the RMSE of ETMonitor vs. EC (MOD16 vs. EC) is 0.59 (1.74) mm/day, 0.39 (0.83) mm/day, 0.39 (0.70) mm/day at the site of Yingke, A'rou, and Guantan, respectively (Figure 6).

5.1.2. Spatial Intercomparison with Precipitation and MOD16 ETa

To reveal the spatial pattern of the ETa from the ETMonitor in relation to the spatial patterns of the relevant environmental and biophysical variables, the annual ETMonitor ETa over the Heihe River basin in 2010 was compared with the TRMM precipitation and MOD16 ETa, respectively (Figure 7). The annual precipitation exhibits a distinguishable southeast to northwest decreasing gradient, corresponding to the climate conditions and topographic altitudes (Figure 7a). The difference between precipitation and ETMonitor ETa indicates that precipitation is higher than ETMonitor ETa in the up reaches of the Heihe River basin due to the high precipitation and low-to-moderate temperatures (Figure 7b). Contrarily, for the oases in the middle and lower reaches, especially for the agriculturally dominated regions whose natural climate alone cannot sustain the current level of agriculture, ETa exceeds precipitation due to importing water by diverting surface water from the Heihe River and reservoirs, or by pumping groundwater, and the results of this comparison are consistent with the regional hydrologic cycle of the basin.

The annual MOD16 ETa in 2010 is about 400–600 mm/year in the up reaches, whereas much lower than 400 mm/year over the irrigated oases in the middle, and even lower (less than 100 mm/year) in the lower reaches of the basin (Figure 7c). Accordingly, the difference between MOD16 and ETMonitor ETa shows very large discrepancy between the up reaches and the middle and lower reaches of the basin (Figure 7d). The spatial intercomparison results between the MOD16 and the ETMonitor ETa indicate that MOD16 ETa is greatly overestimated in the semi-humid up reaches of the Heihe River basin, which agrees with the evaluation results obtained using a water balance method for the upper Yellow River and Yangtze River basins on the Tibetan Plateau [75].

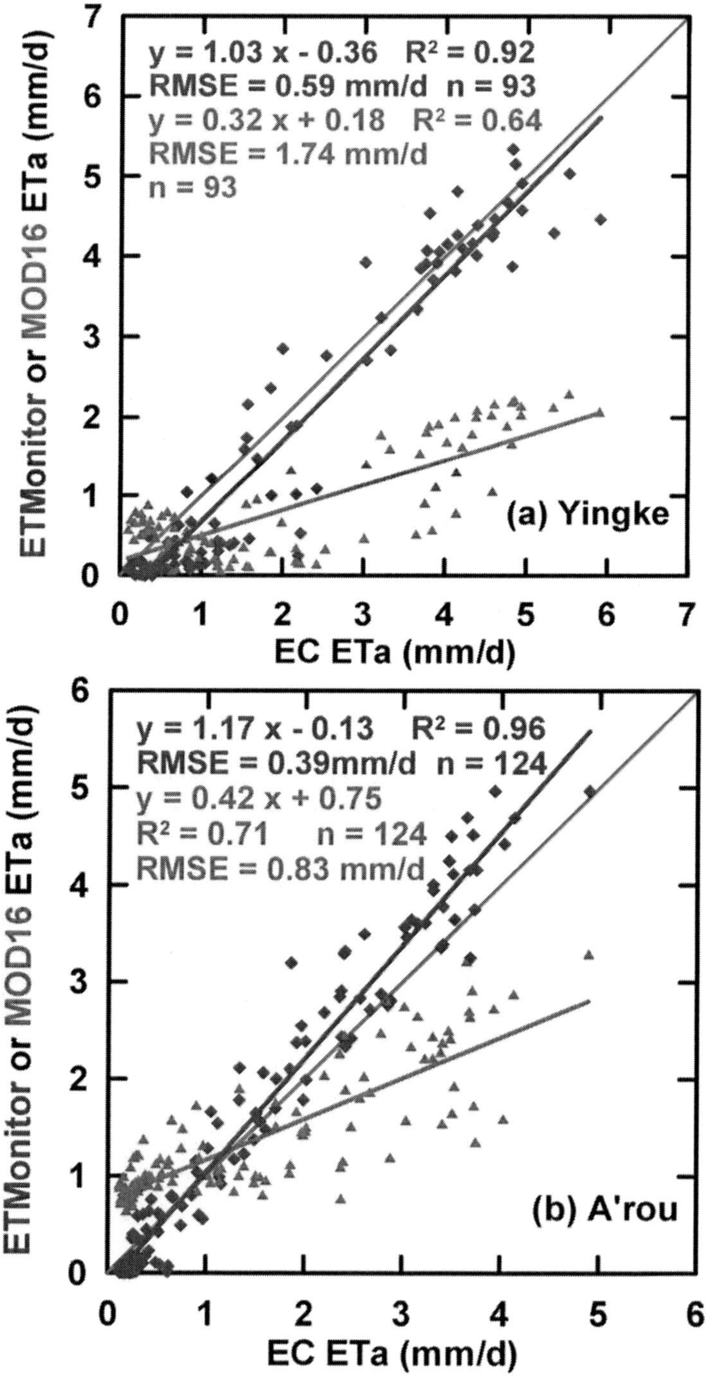
y = 1.03 x - 0.36 R² = 0.92
RMSE = 0.59 mm/d n = 93
y = 0.32 x + 0.18 R² = 0.64
RMSE = 1.74 mm/d
n = 93
(a) Yingke
ETMonitor or MOD16 ETa (mm/d)
EC ETa (mm/d)
y = 1.17 x - 0.13 R² = 0.96
RMSE = 0.39mm/d n = 124
y = 0.42 x + 0.75
R² = 0.71 n = 124
RMSE = 0.83 mm/d
(b) A'rou
ETMonitor or MOD16 ETa (mm/d)
EC ETa (mm/d)

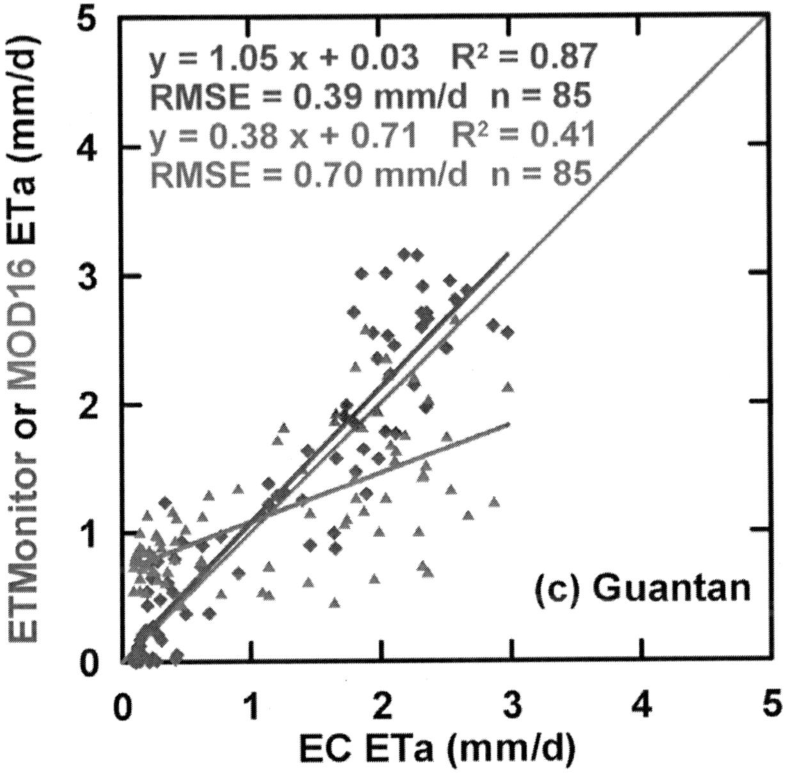

Figure 6. Scattering plots of the ETMonitor (blue diamonds) and MOD16 (green triangles) ETa against EC observations spanning 2009–2011 at the site of (**a**) Yingke (cropland), (**b**) A'rou (alpine meadow), and (**c**) Guantan (forest). Values shown are the eight-day means of ETa (mm/day).

5.2. Sensitivity Analysis

A sensitivity analysis was performed to explore the influence of the environmental and land surface biophysical parameters on the ETa estimates by the ETMonitor model. These parameters include the remote sensing based parameters, the meteorological forcing variables, and the static model parameters. The sensitivity S_p of ETa to a parameter p is defined as [76]:

$$S_{\mathrm{p}} = \left| \frac{ET_{0.9\mathrm{r}} - ET_{1.1\mathrm{r}}}{ET_{\mathrm{r}}} \right| \tag{28}$$

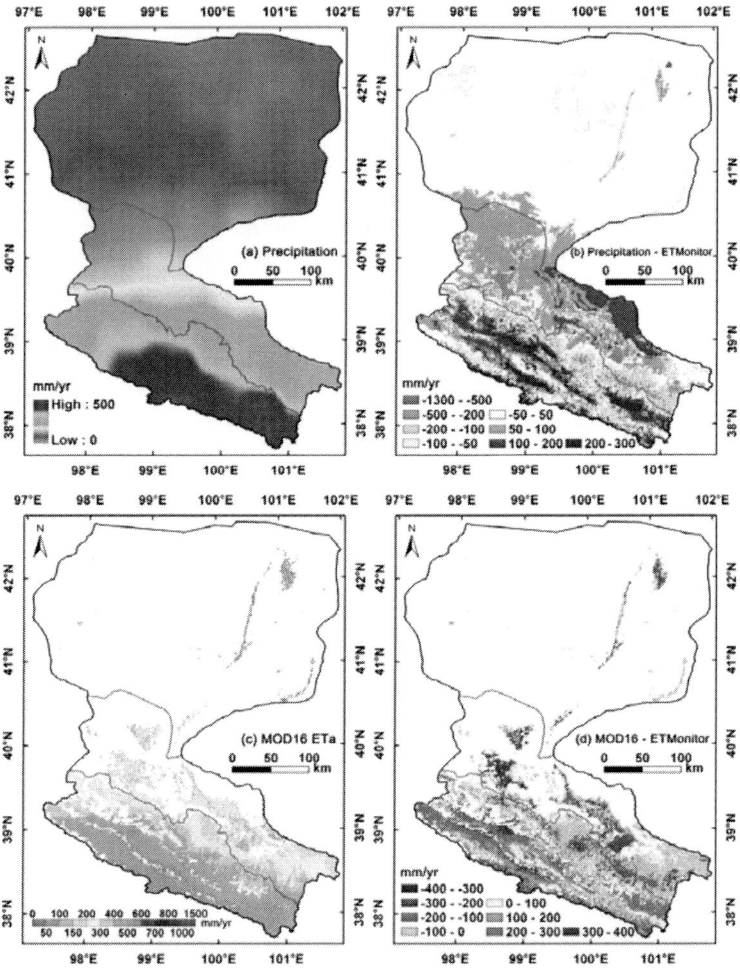

Figure 7. Spatial comparisons of the annual ETMonitor evapotranspiration (ETa) in 2010 with (**a**),(**b**) Tropical Rainfall Measuring Mission (TRMM) precipitation and (**c**),(**d**) MOD16 ETa in the Heihe River basin. In (c) and (d), the regions in white are sparsely vegetated and non-vegetated areas.

where ET_r, $ET_{0.9r}$, and $ET_{1.1r}$ are the ETa (soil evaporation and vegetation canopy transpiration) derived by the ETMonitor when the parameter p equals to its reference value p_r, $0.9p_r$, and $1.1p_r$, respectively, and all other parameters are set to their reference values. The reference values of the remote sensing based parameters (i.e., NDVI, albedo, and surface layer soil water content) and the meteorological forcing variables are taken as their mean values in the growing season (from May to September) at the Yingke site. The sensitivity results are listed in Table 2, which indicate that the model shows larger sensitivity to the

NDVI and the surface layer soil water content among the remote sensing based parameters, to the downward radiation fluxes among the meteorological forcing data, and to the minimum stomatal resistance and the saturated soil water content among the static model parameters. These parameters/variables are the most critical data/parameters needed for further improvement and refinement. It should be noted that the ETMonitor shows much less sensitivity to the wind speed (which is used to calculate the aerodynamic resistance), this can be attributed to the presence of the aerodynamic resistance in both the numerator and in the denominator of the Penman–Monteith equation.

Table 2. Sensitivity of the ETMonitor to environmental and land surface biophysical parameters with ±10% range of variation in the parameters from their reference values.

Parameter		Reference Value	ET_r ($mm \cdot d^{-1}$)	$ET_{0.9r}$ ($mm \cdot d^{-1}$)	$ET_{1.1r}$ ($mm \cdot d^{-1}$)	Sensitivity
NDVI	NDVI (-)	0.66	3.562	2.914	3.964	0.295
Albedo	Albedo (-)	0.19		3.619	3.505	0.032
Surface layer soil water content	θ_g ($cm^3 \cdot cm^{-3}$)	0.12		3.373	3.748	0.105
Air temperature	T_a (°C)	21.3		3.365	3.662	0.084
Air pressure	P (Pa)	84540		3.877	3.287	0.166
Specific humidity	Q ($kg \cdot kg^{-1}$)	0.006		3.605	3.514	0.026
Wind speed	u ($m \cdot s^{-1}$)	3		3.570	3.564	0.002
Downward short-wave radiation	R_s ($W \cdot m^{-2}$)	293.8		3.193	3.932	0.207
Downward long-wave radiation	R_l ($W \cdot m^{-2}$)	330.8		3.230	3.893	0.186
Minimum stomatal resistance	$r_{st,min}$ ($s \cdot m^{-1}$)	180		3.744	3.398	0.097
Minimum soil surface resistance	$r_{s,s,min}$ ($s \cdot m^{-1}$)	50		3.601	3.528	0.021
Saturated soil water content	θ_{sat} ($cm^3 \cdot cm^{-3}$)	0.5		3.768	3.390	0.106
Solar radiation factor	k_{Rs} ($W \cdot m^{-2}$)	500		3.693	3.440	0.071
Optimum air temperature	T_{opt} (°C)	25		3.595	3.510	0.024
Vapor pressure deficit factor	k_{VPD} (hPa^{-1})	0.023		3.671	3.446	0.063
Tenacity factor	K_{sf} (-)	1.5		3.402	3.709	0.086

5.3. ETa Spatial and Temporal Variations in the Heihe River Basin

The multi-year average of annual precipitation for the entire Heihe River basin is 121.9 mm/year. In 2009–2011, the annual precipitation is, respectively, 124.1, 148.5, and 131.0 mm/year, so 2010 is a typical wet year and the estimated ETa of 2010 was primarily used to analyze the spatial and temporal variations of ETa in the Heihe River basin.

5.3.1. Characteristics of ETa in the Upstream, Middle Stream and Downstream Areas

The spatial patterns of the annual ETa spanning 2009–2011 in the Heihe River basin are presented in Figure 8. As expected, the estimated ETa shows consistent patterns with the landscapes over the upstream, middle stream and downstream regions:

(1) In the up reaches of the Heihe River basin, the annual ETa is about 300–600 mm/year in the east part of the up reaches due to the higher vegetation coverage and more humid climate. The annual ETa is less than 300 mm/year in the cold and arid west part of the up reaches. For the regions covered by permanent snow and glaciers in the mountain cryosphere, the magnitude of annual sublimation is about 200–300 mm/year. The seasonal snow is distributed in the mountain cryosphere of the east part of the up reaches, and the snow has completely melted away in summer, resulting that the annual sublimation is much lower than that of the permanent snow and glaciers in the west part of the up reaches.

(2) In the middle reaches, the typical landscapes include irrigated oasis, riparian ecosystem, wetland, and desert. The annual ETa is more than 500 mm/year in the artificial oasis due to irrigation. The major crops are maize, spring wheat and vegetables, which are irrigated about five times during the growing season, and the irrigation quota of the Zhangye Oasis is about 6000 m^3/hm^2 per year (6 ML/ha per year, or 600 mm/year). In the surrounding barren regions the annual ETa is less than 200 mm/year.

(3) In the lower reaches, the annual ETa is about 100–400 mm/year over the river greenbelt and Ejin Oasis and less than 100 mm/year in the most desolate parts of the Gobi desert. The maximum ETa of 1400 mm/year is observed at the terminal East Juyanhai Lake, which has not been dried up since 2004 due to the transfer of water from the Heihe River. The Tian'e Lake, which is located in the southeast of the Ejin Oasis, is seasonally covered by shallow water with the annual ETa of about 700 mm/year in 2010.

The spatial average of the annual ETa in 2009, 2010, and 2011 for the entire Heihe River basin is 140.3, 146.5, and 131.5 mm/year, respectively, which are higher than the multi-year average of 123.3 mm/year derived by the catchment water balance approach [77]. For the catchment water balance approach in [77], the annual evapotranspiration integrated over the entire inland Heihe River basin was estimated as the sum of total annual precipitation, total annual overdraft of groundwater and total annual runoff from the glaciers.

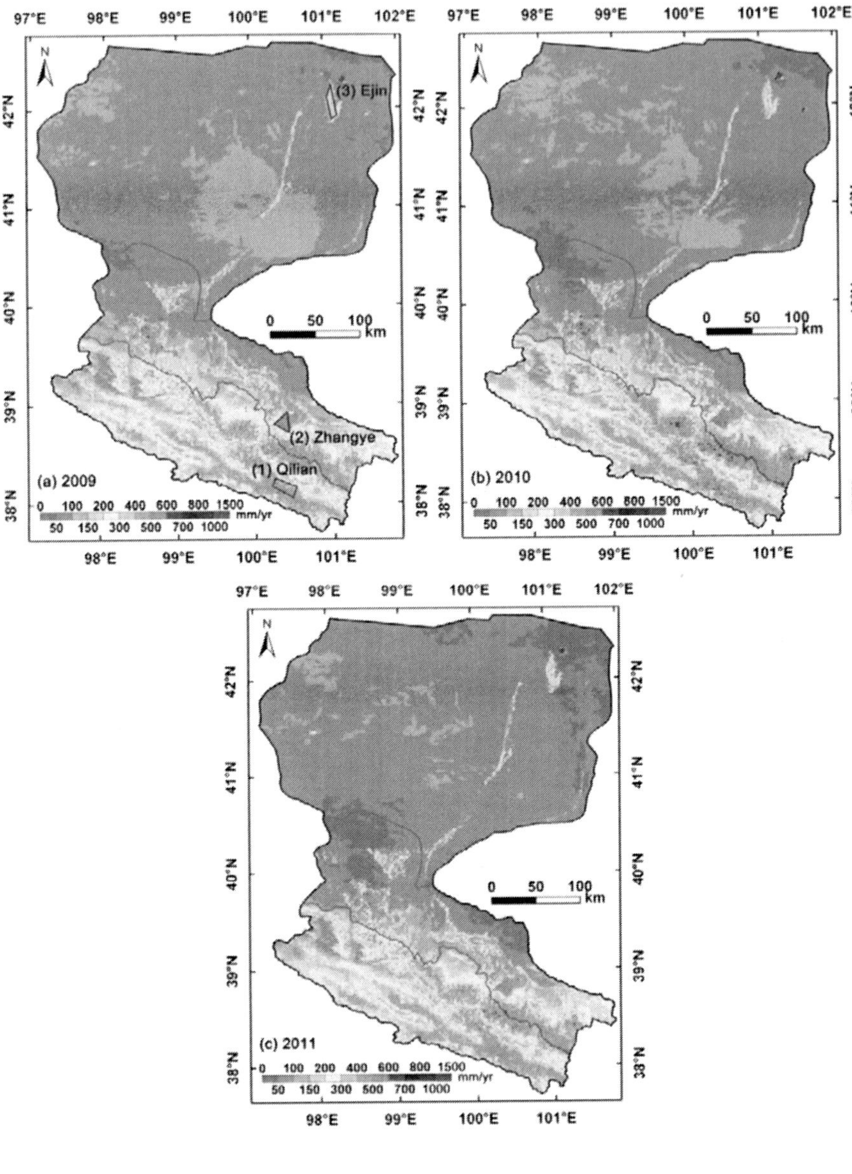

Figure 8. Spatial patterns of the annual ETa spanning 2009–2011 (**a**–**c**) in the Heihe River basin with the maximum ETa of 1400 mm/year at the terminal lake.

The histograms of the annual ETa in the upstream (Qilian), middle stream (Zhangye) and downstream (Ejin) regions indicated by rectangles/triangles in Figure 8a are presented in Figure 9. There are three peak values around 300–350, 400–450, and 500–550 mm/year in the upstream Qilian area (Figure 9a), indicating the strong spatial heterogeneity due to the rolling topography. Whereas in the middle stream and downstream regions, the annual ETa exhibits some homogeneity in the artificial oasis with respect to the upstream area. The annual ETa in 2010 is obviously higher than that in 2009 and 2011 as shown in Figure 9. The annual ETa of the Ejin Oasis is much lower and the histograms in different years are also very similar.

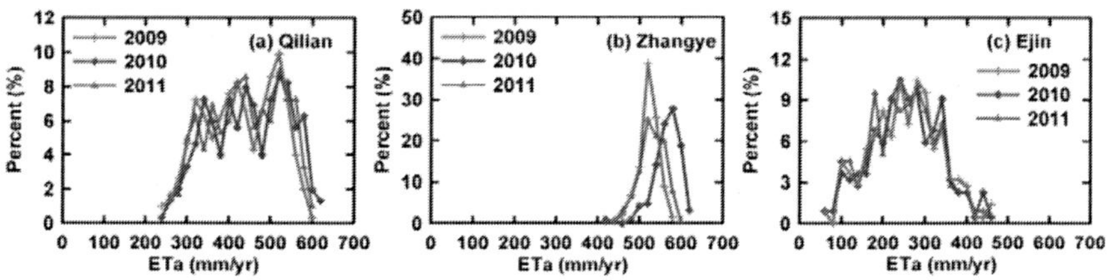

Figure 9. Histograms of the annual ETa in the (**a**) upstream (Qilian), (**b**) middle stream (Zhangye), and (**c**) downstream (Ejin) regions illustrated in Figure 8a.

5.3.2. Land Cover Dependence and Temporal Variation

Evapotranspiration varies substantially between different land cover types in the Heihe River basin, which is demonstrated in Figure 10. The ETa of spruce forest in 2010 is the lowest among the four vegetation types shown in Figure 10a. The potato is mainly distributed near the Jiuquan Oasis and planted in mid-April, whereas the maize is mainly distributed near the Zhangye Oasis and planted in mid-May. The ETa of potato reaches a maximum at around Julian Day of Year (DOY) 169, which is earlier than maize (around DOY 193) and alpine meadow (around DOY 201). At DOY 177, the decrease of ETa is due to the cloudiness, especially for the alpine meadow and spruce forest, which are distributed in the upstream Qilian Mountains of the Heihe River basin.

The sublimation of mountain glaciers distributed in the southwest of Jiuquan varies between 0 and 1.5 mm/day in different seasons of 2010, with the minimum recorded in the period DOY 177–209 due to the decrease of wind speed (Figure 10b). In winter, sublimation is the main kind of water loss over the frozen East Juyanhai Lake. With the increase of air temperature, the ETa of the East Juyanhai Lake gradually increases and reaches a maximum of about 8 mm/day due to the extremely arid climate. The ETa of the Gobi desert in the downstream area of the basin is much less than the ETa of the other land cover types in the Heihe River basin, with the values less than 1 mm/day throughout the year.

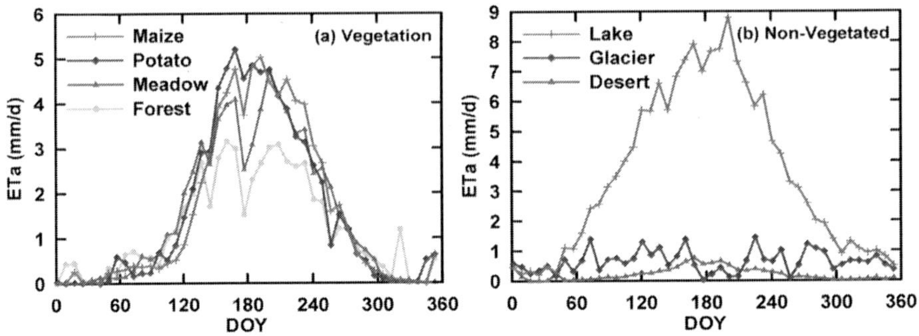

Figure 10. Seasonal variations of the ETa for (**a**) different vegetation types and (**b**) non-vegetated classes in 2010. Values shown are the eight-day means of ETa (mm/day).

The annual ETa of different land cover types in the Heihe River basin is demonstrated in Figure 11. The mean annual ETa of the croplands (maize and potato) is 550–600 mm/year in 2009–2011, a little higher than that of the alpine meadow (550 mm/year) and of the spruce forest (400–450 mm/year). Figure 11 also shows that the highest mean annual ETa values (1200–1400 mm/year) in the Heihe River basin occurred over the East Juyanhai Lake, and the lowest ETa values (less than 100 mm/year) was in the Gobi desert. Besides, the mean sublimation of mountain glaciers is about 200–250 mm/year in 2009–2011.

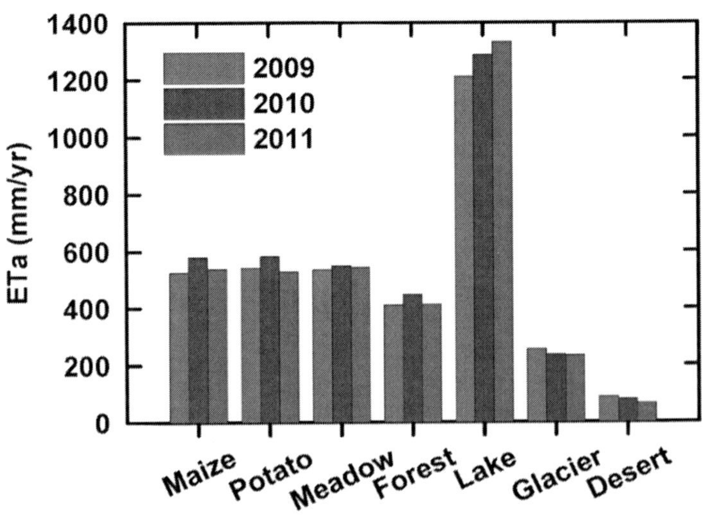

Figure 11. The annual ETa of different land cover types spanning 2009–2011 in the Heihe River basin.

6. DISCUSSION

ETMonitor is developed based on the concept of the Shuttleworth–Wallace dual-source model with some resistance-based methods. The sensitivity analysis indicates that minimizing the sensitivity of the model to wind speed has significant advantages because the wind speed is not easily and accurately determined for heterogeneous landscapes over large scales. While the equations used in the ETMonitor model is physically robust enough, yet they can be solved without any iterative procedure. As such, the ETMonitor is easy to be implemented to generate daily time series of continental and global ETa datasets with a moderate spatial resolution (1 km).

The near-surface meteorological forcing data used in this study was derived based on the Weather Research and Forecasting model (WRF). With respect to data sparse regions, reanalysis meteorological products can be used due to the lack of spatially and temporally complete in-situ observations. The typically used reanalysis products include NCEP–NCAR reanalysis and Climate Forecast System Reanalysis (CFSR) from NCEP; Modern-Era Retrospective Analysis for Research and Applications (MERRA) and Global Land Data Assimilation System (GLDAS) from the Goddard Space Flight Center (GSFC); and 40-yr European Centre for Medium-Range Weather Forecasts (ECMWF) Re-Analysis (ERA-40) and ECMWF Interim Re-Analysis (ERA-Interim) from ECMWF. Decker et al. [78] evaluated these reanalysis products using flux tower observations, concluding that the ERA-Interim dataset is slightly better than all of the rest.

The Qilian Mountain area has a steep topography, and the elevation, aspect and slope are the main topographic factors that affect the distribution of precipitation, temperature, land cover/land use, and soil types in mountain areas indirectly. Topographic influences (elevation, aspect and slope) are important considerations to downscale the model input data in areas with high-localized relative relief [79,80]. As stated in Section 5.1.1, the accuracy of the ETa estimates for the cropland and grassland are higher than that of the forests. The mountain climate has a strong spatial heterogeneity over the mountain area, which cannot be fully captured by the downscaled meteorological forcing data. The complex canopies with short vegetation understories may also lead to the erroneous ETa estimates of the forests. The biases between the ETa estimates and the EC measurements can also arise from the inaccuracy of the EC measurements which have an uncertainty of about 10%–30% [19], especially for the areas with rolling topography. Besides, it should be noted that the following aspects need further investigations and studies:

(1) During different stages of the vegetation growing season from the initial, through development and mid-season, and to the late season, the minimum stomatal resistance $r_{st,min}$ may vary substantially between the four distinct growth stages of various types of vegetation. In ETMonitor, $r_{st,min}$ is taken as a

single constant value for each vegetation type spanning the entire growing season, which needs to be improved based on the field observations in the Heihe River basin.

(2) The soil moisture data derived from passive microwave sensors, such as AMSR-E, Soil Moisture and Ocean Salinity (SMOS), and Fengyun-3 (FY-3) MicroWave Radiation Imager (MWRI) have a coarse spatial resolution (25 km), which is not sufficient for regional scale studies [73,74]. The accuracy and spatial resolution will improve in the near future [81,82]. The on-going National Aeronautics and Space Administration (NASA) Soil Moisture Active and Passive (SMAP) mission (launched on 31 January 2015) will provide more precise soil moisture data at a higher spatial resolution [83]. The spatial resolution of SMAP Level 3 soil moisture data product is 3 km (Radar), 36 km (Radiometer), and 9 km (Radar + Radiometer). The SMAP Level 4 product will provide a global surface and root zone soil moisture data set at a spatial resolution of 9 km. Besides, some downscaling methods to downscale the spatial resolution of passive microwave soil moisture data have achieved some developments [84], and a soil moisture product with a spatial resolution of 1 km will be generated by merging microwave remote sensing data and MODIS VI products. This is very useful to develop and improve the performance of ETa models based on soil moisture data.

(3) The spatial and temporal resolutions of the TRMM precipitation data used in this study is 0.25°/tri-hourly, which is also not sufficient for regional scale studies. The Global Precipitation Measurement (GPM) mission has been designed to provide more accurate/frequent observations of global precipitation using the GPM Core Observatory satellite (launched on 27 February 2014 by NASA-JAXA) with the Constellation Satellites (international partners' satellites with precipitation measurement capability). The 0.1°/30-minute global (90°N–90°S) precipitation data have been released in near real time (about 4 hours after observation) since March 2014 to present (available at http://pmm.nasa.gov/data-access/downloads/gpm (accessed on 15 January 2015)). Such data are very useful to improve the accuracy of the canopy interception estimates.

As stated in Section 5.1.1, the MOD16 ETa is underestimated in the growing season, especially for the irrigated artificial oasis located in a semi-arid climate. Such findings are in accordance with those obtained from previous studies over China [85,86], the conterminous United States [87], Brazil [88], South Africa [89], and Spain [90]. MOD16 ETa is not constrained explicitly by soil moisture and assumes that water availability can be captured by atmospheric VPD [91,92]. For soil evaporation estimation, the formulation of the soil moisture constraint in the MOD16 ETa algorithm does not include the stress factor for surface layer soil moisture, whereas the VPD and relative humidity are used as indicators of water stress. For canopy transpiration estimation, the formulation of canopy resistance in the MOD16 ETa algorithm is based on the VPD rather than the root-zone soil moisture. When the soil water content in the surface and root-zone layers are high, e.g., due to irrigation or thawing of frozen soil, the soil evaporation and canopy transpiration are underestimated due to the overestimation of water stress parameterized by the atmospheric VPD and relative humidity in the semi-arid middle reaches of the Heihe River basin.

7. CONCLUSIONS

We developed a remotely sensed ETa model (ETMonitor) to estimate the daily actual evapotranspiration of the Heihe River basin for the years 2009–2011 at a spatial resolution of 1 km. The estimated ETa was compared with the annual TRMM precipitation and the MOD16 ETa product, and was evaluated against in-situ measurements at the site of Yingke (cropland), A'rou (alpine meadow), and Guantan (forest), respectively. The evaluation and validation results demonstrate that the ETMonitor model performs better for alpine meadow (grassland) and cropland than for the forest, and achieves much higher accuracy compared with the MOD16 ETa. The soil moisture is a major factor contributing to ETa underestimation for cropland in spring, which needs to be improved both in the aspect of retrieval algorithm and downscaling.

The estimated ETa shows reasonable spatial and temporal patterns with respect to the diverse cold and arid landscapes in the Heihe River basin. The terminal lake surrounded by desert has the highest ETa with the annual ETa of 1400 mm/year in 2010. The ETa of crops in the irrigated artificial oasis is much higher than the natural vegetation, e.g., alpine meadow and spruce forest. The seasonal variation of ETa for a specific crop variety is primarily determined by its phenology. The daily ETa is produced for all sky conditions, which is continuous and consistent at spatial and temporal scales and useful for integrated eco-hydrological studies and water resource management to improve the rational allocation of water resources with more efficiency, equity, and sustainability in the Heihe River basin. Registered users can access these data (2009–2011) through the Heihe Plan Science Data Center (http://www.heihedata.org/, doi:10.3972/heihe.114.2013.db (accessed on 15 January 2015)). Long-term series of ETa will be produced covering the years 2001–2015 at 1 km and daily resolutions, which will be used for the prediction and estimation of regional-scale surface runoff and groundwater, for the scheduling of field irrigations, and for the validation of the integrated water–ecosystem–economy model developed for the Heihe River basin [93].

ACKNOWLEDGMENTS

This work is supported by the National Natural Science Foundation of China (NSFC) (Grant no. 91425303) and the National Key Basic Research Program of China (Grant no. 2015CB953702). Eddy covariance flux tower sites are collected from Watershed Allied Telemetry Experimental Research (WATER) and provided by Cold and Arid Regions Sciences Data Center at Lanzhou, and we gratefully acknowledge the efforts of researchers at these sites.

REFERENCES

1. Nian, Y.Y.; Li, X.; Zhou, J.; Hu, X.L. Impact of land use change on water resource allocation in the middle reaches of the Heihe River Basin in northwestern China. J. Arid Land **2014**, 6, 273–286.
2. Zeng, Z.; Liu, J.; Koeneman, P.H.; Zarate, E.; Hoekstra, A.Y. Assessing water footprint at river basin level: A case study for the Heihe River Basin in northwest China. Hydrol. Earth Syst. Sci. **2012**, 16, 2771–2781.
3. Ge, Y.C.; Li, X.; Huang, C.L.; Nan, Z.T. A Decision Support System for irrigation water allocation along the middle reaches of the Heihe River Basin, Northwest China. Environ. Model. Softw. **2013**, 47, 182–192.
4. Leuning, R.; Zhang, Y.Q.; Rajaud, A.; Cleugh, H.; Tu, K. A simple surface conductance model to estimate regional evaporation using MODIS leaf area index and the Penman–Monteith equation. Water Resour. Res. **2008**.
5. Carlson, T.N.; Capehart, W.J.; Gillies, R.R. A new look at the simplified method for remote sensing of daily evapotranspiration. Remote Sens. Environ. **1995**, 54, 161–167.
6. Wang, K.C.; Liang, S.L. An improved method for estimating global evapotranspiration based on satellite determination of surface net radiation, vegetation index, temperature, and soil moisture. J. Hydrometeorol. **2008**, 9, 712–727.
7. Garcia, M.; Fernández, N.; Villagarcía, L.; Domingo, F.; Puigdefábregas, J.; Sandholt, I. Accuracy of the Temperature-Vegetation Dryness Index using MODIS under water-limited vs. energy-limited evapotranspiration conditions. Remote Sens. Environ. **2014**, 149, 100–117.
8. Jiang, L.; Islam, S. A methodology for estimation of surface evapotranspiration over large areas using remote sensing observations. Geophys. Res. Lett. **1999**, 26, 2773–2776.
9. Tang, R.L.; Li, Z.L.; Tang, B.H. An application of the Ts–VI triangle method with enhanced edges determination for evapotranspiration estimation from MODIS data in arid and semi-arid regions: Implementation and validation. Remote Sens. Environ. **2010**, 114, 540–551.

10. Bastiaanssen, W.G.M.; Menenti, M.; Feddes, R.A.; Holtslag, A.A.M. A remote sensing surface energy balance algorithm for land (SEBAL): 1. Formulation. J. Hydrol. **1998**, 212, 198–212.
11. Jia, L.; Su, Z.; van den Hurk, B.; Menenti, M.; Moene, A.; De Bruin, H.A.R.; Yrisarry, J.J.B.; Ibanez, M.; Cuesta, A. Estimation of sensible heat flux using the Surface Energy Balance System (SEBS) and ATSR measurements. Phys. Chem. Earth **2003**, 28, 75–88.
12. Kustas, W.P.; Norman, J.M. Evaluation of soil and vegetation heat flux predictions using a simple two-source model with radiometric temperatures for partial canopy cover. Agric. For. Meteorol. **1999**, 94, 13–29.
13. Norman, J.M.; Kustas, W.P.; Humes, K.S. Source approach for estimating soil and vegetation energy fluxes in observations of directional radiometric surface temperature. Agric. For. Meteorol. **1995**, 77, 263–293.
14. Su, Z. The Surface Energy Balance System (SEBS) for estimation of turbulent heat fluxes. Hydrol. Earth Syst. Sci. **2002**, 6, 85–99.
15. Bastiaanssen, W.G.M.; Cheema, M.J.M.; Immerzeel, W.W.; Miltenburg, I.J.; Pelgrum, H. Surface energy balance and actual evapotranspiration of the transboundary Indus Basin estimated from satellite measurements and the ETLook model. Water Resour. Res. **2012**.
16. Cleugh, H.A.; Leuning, R.; Mu, Q.; Running, S.W. Regional evaporation estimates from flux tower and MODIS satellite data. Remote Sens. Environ. **2007**, 106, 285–304.
17. Fisher, J.B.; Tu, K.P.; Baldocchi, D.D. Global estimates of the land atmosphere water flux based on monthly AVHRR and ISLSCP-II data, validated at FLUXNET sites. Remote Sens. Environ. **2008**, 112, 901–919.
18. Mu, Q.Z.; Heinsch, F.A.; Zhao, M.S.; Running, S.W. Development of a global evapotranspiration algorithm based on MODIS and global meteorology data. Remote Sens. Environ. **2007**, 111, 519–536.
19. Mu, Q.Z.; Zhao, M.S.; Running, S.W. Improvements to a MODIS global terrestrial evapotranspiration algorithm. Remote Sens. Environ. **2011**, 115, 1781–1800.
20. Courault, D.; Seguin, B.; Olioso, A. Review on estimation of evapotranspiration from remote sensing data: From empirical to numerical modeling approaches. Irrig. Drain. Syst. **2005**, 19, 223–249.

21. Kalma, J.D.; McVicar, T.R.; McCabe, M.F. Estimating land surface evaporation: A review of methods using remotely sensed surface temperature data. Surv. Geophys. **2008**, 29, 421–469.
22. Li, Z.L.; Tang, R.L.; Wan, Z.; Bi, Y.; Zhou, C.; Tang, B.H.; Yan, G.J.; Zhang, X. A review of current methodologies for regional evapotranspiration estimation from remotely sensed data. Sensors **2009**, 9, 3801–3853.
23. Li, S.B.; Zhao, W.Z. Satellite-based actual evapotranspiration estimation in the middle reach of the Heihe River Basin using the SEBAL method. Hydrol. Process. **2010**, 24, 3337–3344.
24. Yang, Y.M.; Su, H.B.; Zhang, R.H.; Tian, J.; Yang, S.Q. Estimation of regional evapotranspiration based on remote sensing: Case study in the Heihe River Basin. J. Appl. Remote Sens. **2012**.
25. Li, Z.L.; Tang, B.H.; Wu, H.; Ren, H.Z.; Yan, G.J.; Wan, Z.M.; Trigo, I.F.; Sobrino, J.A. Satellite-derived land surface temperature: Current status and perspectives. Remote Sens. Environ. **2013**, 131, 14–37.
26. Ghilain, N.; Arboleda, A.; Sepulcre-Canto, G.; Batelaan, O.; Ardo, J.; Gellens-Meulenberghs, F. Improving evapotranspiration in a land surface model using biophysical variables derived from MSG/SEVIRI satellite. Hydrol. Earth Syst. Sci. **2012**, 16, 2567–2583.
27. Monteith, J.L. Evaporation and environment. In The State and Movement of Water in Living Organisms. Symposium of the Society of Experimental Biology; Fogg, G.E., Ed.; Society of Experimental Biology, Cambridge University Press: Swansea, UK, 1965; Volume 19, pp. 205–234.
28. Liang, X.; Lettenmaier, D.P.; Wood, E.F.; Burges, S.J. A simple hydrologically based model of land surface water and energy fluxes for general circulation models. J. Geophys. Res. **1994**, 99, 14415–14428.
29. Sellers, P.J.; Randall, D.A.; Collatz, G.J.; Berry, J.A.; Field, C.B.; Dazlich, D.A.; Zhang, C.; Collelo, G.D.; Bounoua, L. A revised land surface parameterization (SiB2) for atmospheric GCMs. Part I: Model formulation. J. Clim. **1996**, 9, 676–705. [Google Scholar] [CrossRef]
30. Shuttleworth, W.J.; Wallace, J.S. Evaporation from sparse crops—An energy combination theory. Q. J. R. Meteorol. Soc. **1985**, 111, 839–855.
31. Anadranistakis, M.; Liakatas, A.; Kerkides, P.; Rizosb, S.; Gavanosisb, J.; Poulovassilis, A. Crop water requirements model tested for crops grown in Greece. Agric. Water Manag. **2000**, 45, 297–316.
32. Shaw, R.H.; Pereira, A.R. Aerodynamic roughness of a plant canopy: A numerical experiment. Agric. Meteorol. **1982**, 26, 51–65.

33. Gutman, G.; Ignatov, V. The derivation of the green vegetation fraction from NOAA/AVHRR data for use in numerical weather prediction models. Int. J. Remote Sens. **1998**, 19, 1533–1543.
34. Camillo, P.J.; Gurney, R.J. A resistance parameter for bare-soil evaporation models. Soil Sci. **1986**, 141, 95–105.
35. Clapp, R.B.; Hornberger, G.M. Empirical equations for some soil hydraulic properties. Water Resour. Res. **1978**, 14, 601–604.
36. Dolman, A.J. A multiple-source land surface energy balance model for use in general circulation models. Agric. For. Meteorol. **1993**, 65, 21–45.
37. Van Genuchten, M.T. A closed-form equation for predicting the hydraulic conductivity of unsaturated soils. Soil Sci. Soc. Am. J. **1980**, 44, 892–898.
38. Avissar, R.; Avissar, P.; Mahrer, Y.; Bravdo, B.A. A model to simulate response of plant stomata to environmental conditions. Agric. For. Meteorol. **1985**, 34, 21–29.
39. Jarvis, P.G. The interpretation of the variations in leaf water potential and stomatal conductance found in canopies in the field. Philos. Trans. R. Soc. Lond. B **1976**, 273, 593–610.
40. Stewart, J.B. Modelling surface conductance of pine forest. Agric. For. Meteorol. **1988**, 43, 19–35.
41. Mehrez, M.B.; Taconet, O.; Vidal-Madjar, D.; Valencogne, C. Estimation of stomatal resistance and canopy evaporation during the HAPEX–MOBILHY experiment. Agric. For. Meteorol. **1992**, 58, 285–313.
42. Dickinson, R.E. Modelling evapotranspiration for three-dimensional global climate models. Geophys. Monogr. Ser. **1984**, 29, 58–72.
43. Irannejad, P.; Shao, Y.P. Description and validation of the atmosphere–land–surface interaction scheme (ALSIS) with HAPEX and Cabauw data. Glob. Planet. Chang. **1998**, 19, 87–114.
44. Mo, X.G.; Liu, S.X.; Lin, Z.H.; Zhao, W.M. Simulating temporal and spatial variation of evapotranspiration over the Lushi basin. J. Hydrol. **2004**, 285, 125–142.
45. Sellers, P.J.; Mintz, Y.; Sud, Y.C.; Dalcher, A. A simple biosphere model (SiB) for use within general circulation models. J. Atmos. Sci. **1986**, 43, 505–531. [Google Scholar] [CrossRef]
46. Noilhan, J.; Planton, S. A simple parameterization of land surface processes for meteorological models. Mon. Weather Rev. **1989**, 117, 536–549.

47. Allen, R.G.; Pruitt, W.O.; Businger, J.A.; Fritschen, L.J.; Jensen, M.E.; Quinn, F.H. Evaporation and transpiration. In Hydrology Handbook, 2nd ed.; Task Committee on Hydrology Handbook of Management Group D of American Society of Civil Engineers (ASCE), Ed.; ASCE: New York, NY, USA, 1996; pp. 125–252.
48. Gash, J.H.C.; Lloyd, C.R.; Lachaud, G. Estimating sparse forest rainfall interception with an analytical model. J. Hydrol. **1995**, 170, 79–86.
49. Van Dijk, A.I.J.M.; Bruijnzeel, L.A. Modelling rainfall interception by vegetation of variable density using an adapted analytical model. Part 1. Model description. J. Hydrol. **2001**, 247, 230–238.
50. Cui, Y.K.; Jia, L. A modified Gash model for estimating rainfall interception loss of forest using remote sensing observations at regional scale. Water **2014**, 6, 993–1012.
51. Cui, Y.K.; Jia, L.; Hu, G.C.; Zhou, J. Mapping of interception loss of vegetation in the Heihe River basin of China using remote sensing observations. IEEE Geosci. Remote Sens. Lett. **2015**, 12, 23–27.
52. Kuzmin, P.P. On method for investigations of evaporation from the snow cover. Trans. State Hydrol. Inst. **1953**, 41, 34–52. (In Russian).
53. Penman, H.L. Natural evaporation from open water, bare soil and grass. Proc. R. Soc. A **1948**, 193, 120–145.
54. Liu, Q.; Wang, L.; Qu, Y.; Liu, N.; Liu, S.; Tang, H.; Liang, S. Preliminary evaluation of the long-term GLASS albedo product. Int. J. Digit. Earth **2013**, 6, 69–95.
55. Jia, L.; Shang, H.; Hu, G.; Menenti, M. Phenological response of vegetation to upstream river flow in the Heihe River basin by time series analysis of MODIS data. Hydrol. Earth Syst. Sci. **2011**, 15, 1047–1064.
56. Ran, Y.H.; Li, X.; Lu, L.; Li, Z.Y. Large-scale land cover mapping with the integration of multi-source information based on the Dempster–Shafer theory. Int. J. Geogr. Inf. Sci. **2012**, 26, 169–191.
57. Zhong, B.; Ma, P.; Nie, A.H.; Yang, A.X.; Yao, Y.J.; Lv, W.B.; Zhang, H.; Liu, Q.H. Land cover mapping using time series HJ-1/CCD data. Sci. China Earth Sci. **2014**, 57, 1790–1799.
58. Dai, Y.J.; Shangguan, W.; Duan, Q.Y.; Liu, B.; Fu, S.; Niu, G. Development of a China dataset of soil hydraulic parameters using pedotransfer functions for land surface modeling. J. Hydrometeorol. **2013**, 14, 869–887.
59. Pan, X.D.; Li, X. Validation of WRF model on simulating forcing data for Heihe River Basin. Sci. Cold Arid Region. **2011**, 3, 344–357.

60. Pan, X.D.; Li, X.; Shi, X.K.; Han, X.J.; Luo, L.H.; Wang, L.X. Dynamic downscaling of near-surface air temperature at the basin scale using WRF—A case study in the Heihe River Basin, China. Front. Earth Sci. **2012**, 6, 314–323.
61. Gao, Y.C.; Long, D.; Li, Z.L. Estimation of daily actual evapotranspiration from remotely sensed data under complex terrain over the upper Chao river basin in North China. Int. J. Remote Sens. **2008**, 29, 3295–3315.
62. Stahl, K.; Moore, R.D.; Floyer, J.A.; Asplin, M.G.; McKendry, I.G. Comparison of approaches for spatial interpolation of daily air temperature in a large region with complex topography and highly variable station density. Agric. For. Meteorol. **2006**, 139, 224–236.
63. Li, X.; Li, X.W.; Li, Z.Y.; Ma, M.G.; Wang, J.; Xiao, Q.; Liu, Q.; Che, T.; Chen, E.X.; Yan, G.J.; et al. Watershed allied telemetry experimental research. J. Geophys. Res. **2009**.
64. Li, X.; Cheng, G.D.; Liu, S.M.; Xiao, Q.; Ma, M.G.; Jin, R.; Che, T.; Liu, Q.H.; Wang, W.Z.; Qi, Y.; et al. Heihe Watershed Allied Telemetry Experimental Research (HiWATER) scientific objectives and experimental design. Bull. Am. Meteorol. Soc. **2013**, 94, 1145–1160.
65. Liu, S.M.; Xu, Z.W.; Wang, W.Z.; Jia, Z.Z.; Zhu, M.J.; Bai, J.; Wang, J.M. A comparison of eddy-covariance and large aperture scintillometer measurements with respect to the energy balance closure problem. Hydrol. Earth Syst. Sci. **2011**, 15, 1291–1306.
66. Blanken, P.D.; Black, T.A.; Neumann, H.H.; Den Hartog, G.; Yang, P.C.; Nesic, Z.; Staebler, R.; Chen, W.; Novak, M.D. Turbulent flux measurements above and below the overstory of a boreal aspen forest. Bound. Layer Meteorol. **1998**, 89, 109–140.
67. Goulden, M.L.; Daube, B.C.; Fan, S.M.; Sutton, D.J.; Bazzaz, A.; Munger, J.W.; Wofsy, S.C. Physiological responses of a black spruce forest to weather. J. Geophys. Res. **1997**, 102, 28987–28996.
68. Meyers, T.P.; Hollinger, S.E. An assessment of storage terms in the surface energy balance of maize and soybean. Agric. For. Meteorol. **2004**, 125, 105–115.
69. Twine, T.E.; Kustas, W.P.; Norman, J.M.; Cook, D.R.; Houser, P.R.; Meyers, T.P.; Prueger, J.H.; Starks, P.J.; Wesely, M.L. Correcting eddy-covariance flux underestimates over a grassland. Agric. For. Meteorol. **2000**, 103, 279–300.

70. Kustas, W.P.; Norman, J.M. Evaluating the effects of subpixel heterogeneity on pixel average fluxes. Remote Sens. Environ. **2000**, 74, 327–342.

71. Kustas, W.P.; Li, F.; Jackson, T.J.; Prueger, J.H.; MacPherson, J.I.; Wolde, M. Effects of remote sensing pixel resolution on modeled energy flux variability of croplands in Iowa. Remote Sens. Environ. **2004**, 92, 535–547.

72. Li, F.; Kustas, W.P.; Anderson, M.C.; Prueger, J.H.; Scott, R.L. Effect of remote sensing spatial resolution on interpreting tower-based flux observations. Remote Sens. Environ. **2008**, 112, 337–349.

73. Chen, Y.Y.; Yang, K.; Qin, J.; Zhao, L.; Tang, W.J.; Han, M.L. Evaluation of AMSR-E retrievals and GLDAS simulations against observations of a soil moisture network on the central Tibetan Plateau. J. Geophys. Res. Atmos. **2013**, 118, 4466–4475.

74. Yang, K.; Qin, J.; Zhao, L.; Chen, Y.Y.; Tang, W.J.; Han, M.L.; Lazhu; Chen, Z.Q.; Lv, N.; Ding, B.H.; et al. A multi-scale soil moisture and freeze-thaw monitoring network on the Third Pole. Bull. Am. Meteorol. Soc. **2013**, 94, 1907–1916.

75. Xue, B.L.; Wang, L.; Li, X.P.; Yang, K.; Chen, D.L.; Sun, L.T. Evaluation of evapotranspiration estimates for two river basins on the Tibetan Plateau by a water balance method. J. Hydrol. **2013**, 492, 290–297.

76. Zhan, X.; Kustas, W.P.; Humes, K.S. An intercomparison study on models of sensible heat flux over partial canopy surfaces with remotely sensed surface temperature. Remote Sens. Environ. **1996**, 58, 242–256.

77. Cheng, G.D. Integrated Management of the Water–Ecology–Economy System in the Heihe River Basin; Science Press: Beijing, China, 2009.

78. Decker, M.; Brunke, M.A.; Wang, Z.; Sakaguchi, K.; Zeng, X.B.; Bosilovich, M.G. Evaluation of the reanalysis products from GSFC, NCEP, and ECMWF using flux tower observations. J. Clim. **2012**, 25, 1916–1944.

79. Liston, G.E.; Elder, K. A meteorological distribution system for high-resolution terrestrial modeling (MicroMet). J. Hydrometeorol. **2006**, 7, 217–234.

80. McVicar, T.R.; Van Niel, T.G.; Li, L.T.; Hutchinson, M.F.; Mu, X.M.; Liu, Z.H. Spatially distributing monthly reference evapotranspiration and pan evaporation considering topographic influences. J. Hydrol. **2007**, 338, 196–220.

81. Liu, Q.; Du, J.Y.; Shi, J.C.; Jiang, L.M. Analysis of spatial distribution and multi-year trend of the remotely sensed soil moisture on the Tibetan Plateau. Sci. China Earth Sci. **2013**, 56, 2173–2185.
82. Shi, J.C.; Du, Y.; Du, J.Y.; Jiang, L.M.; Chai, L.N.; Mao, K.B.; Xu, P.; Ni, W.J.; Xiong, C.; Liu, Q.; et al. Progresses on microwave remote sensing of land surface parameters. Sci. China Earth Sci. **2012**, 55, 1052–1078.
83. Das, N.N.; Entekhabi, D.; Njoku, E.G. An algorithm for merging SMAP radiometer and radar data for high-resolution soil-moisture retrieval. IEEE Trans. Geosci. Remote Sens. **2011**, 49, 1504–1512.
84. Song, C.Y.; Jia, L.; Menenti, M. Retrieving high-resolution surface soil moisture by downscaling AMSR-E brightness temperature using MODIS LST and NDVI data. IEEE J. Sel. Top. Appl. Earth Obs. Remote Sens. **2014**, 7, 935–942.
85. Liu, S.M.; Xu, Z.W.; Zhu, Z.L.; Jia, Z.Z.; Zhu, M.J. Measurements of evapotranspiration from eddy-covariance systems and large aperture scintillometers in the Hai River Basin, China. J. Hydrol. **2013**, 487, 24–38.
86. Liu, Z.J.; Shao, Q.Q.; Liu, J.Y. The performances of MODIS-GPP and -ET products in China and their sensitivity to input data (FPAR/LAI). Remote Sens. **2015**, 7, 135–152.
87. Velpuri, N.M.; Senay, G.B.; Singh, R.K.; Bohms, S.; Verdin, J.P. A comprehensive evaluation of two MODIS evapotranspiration products over the conterminous United States: Using point and gridded FLUXNET and water balance ET. Remote Sens. Environ. **2013**, 139, 35–49.
88. Ruhoff, A.L.; Paz, A.R.; Aragao, L.E.O.C.; Mu, Q.; Malhi, Y.; Collischonn, W.; Rocha, H.R.; Running, S.W. Assessment of the MODIS global evapotranspiration algorithm using eddy covariance measurements and hydrological modelling in the Rio Grande basin. Hydrol. Sci. J. **2013**, 58, 1–19.
89. Ramoelo, A.; Majozi, N.; Mathieu, R.; Jovanovic, N.; Nickless, A.; Dzikiti, S. Validation of global evapotranspiration product (MOD16) using flux tower data in the African savanna, South Africa. Remote Sens. **2014**, 6, 7406–7423.
90. Hu, G.C.; Jia, L.; Menenti, M. Comparison of MOD16 and LSA-SAF MSG evapotranspiration products over Europe for 2011. Remote Sens. Environ. **2015**, 156, 510–526.
91. Garcia, M.; Sandholt, I.; Ceccato, P.; Ridler, M.; Mougin, E.; Kergoat, L.; Morillas, L.; Timouk, F.; Fensholt, R.; Domingo, F. Actual

evapotranspiration in drylands derived from in-situ and satellite data: Assessing biophysical constraints. Remote Sens. Environ. **2013**, 131, 103–118.

92. Vinukollu, R.K.; Wood, E.F.; Ferguson, C.R.; Fisher, J.B. Global estimates of evapotranspiration for climate studies using multi-sensor remote sensing data: Evaluation of three process-based approaches. Remote Sens. Environ. **2011**, 115, 801–823.

93. Cheng, G.D.; Li, X.; Zhao, W.Z.; Xu, Z.M.; Feng, Q.; Xiao, S.C.; Xiao, H.L. Integrated study of the water–ecosystem–economy in the Heihe River Basin. Natl. Sci. Rev. **2014**, 1, 413–428.

Chapter 10

Use of Evapotranspiration (ET) Landfill Covers to Reduce Methane Emissions from Municipal Solid Waste Landfills

Tarek Abichou[1*], Tarek Kormi[2], Cheng Wang[3], Haykel Melaouhia[1], Terry Johnson[4], Stephen Dwyer[5]

[1]Department of Civil Engineering, Florida State University, Tallahassee, FL, USA

[2]Ecole Nationale d' Ingenieurs de Gabes, Department de Genie Civil, Gabes, Tunisia, LASMAP, Ecole Polytechnique de Tunisie, University of Carthage, Tunisia

[3]Argonne National Laboratory, Chicago, IL, USA.

[4]Waste Management, Minneapolis, MN, USA.

[5]Dwyer Engineering, Albuquerque, NM, USA

ABSTRACT

Solid waste landfills need to have final covers to 1) reduce the infiltration of rainfall into the waste mass and 2) reduce surface greenhouse gas emissions. Most regulations require that such final covers include hydraulic barriers, such as compacted clays with or without geomembrane. Significant research has been undertaken to allow the use of evapotranspiration-based covers (often termed: Evapotranspiration (ET) Cover, Water Balance Covers, or Phyto Covers) as an alternative to the barrier concept covers. ET covers are designed so that they have the capacity to store water by the soil and also have plants or vegetation to remove the stored water. In ET covers, plant roots can enhance the aeration of soil by creating secondary macropores which improve the diffusion of oxygen into soil. Therefore, biological methane oxidation (a natural process in landfill soils) can be improved considerably by the soil structuring processes of vegetation, along with the increase of organic biomass in the soil associated with plant roots. This paper summarizes a study to investigate the capacity of an ET cover to reduce surface greenhouse gas emissions when implemented on a solid waste landfill. This study consisted of using a numerical model to estimate methane emission and oxidation through an ET cover under average climatic conditions in Bennignton, Nebraska, USA. Different simulations were performed using different methane loading flux (5 to 200 $gm^{-2}\cdot d^{-1}$) as the bottom boundary. For all simulations, surface emissions were the lowest during the growing season and during warmer days of the year. Percent oxidation is the highest during the growing season and during warmer days. The lowest modeled surface emissions were always obtained during the growing season. Finally, correlations between percent oxidation and methane loading into simulated ET covers were proposed to estimate methane emissions and methane oxidation in ET covers.

Keywords: Evapotranspiration Covers, Methane Oxidation, Landfills, Greenhouse Gas Emissions

1. INTRODUCTION

1.1. Background

Final covers are used to reduce the quantity of water that infiltrates into waste deposits at landfills. Reducing the volume of infiltrating water reduces the amount of leachate that is generated and the risk of groundwater contamination. Generally, the applicable rules and regulations require that landfill covers employ resistive principles, i.e., layers having low saturated hydraulic conductivity such as compacted clay barriers, geosynthetic clay liners with or without a geomembrane. This design philosophy is often referred to as "raincoat", barrier, or "umbrella" approach. Barrier type covers, such as compacted clay covers, have been shown to loose their impermeable qualities over time because of the influence of climate variations on the integrity of the soil layer [1]. The United States Environmental Protection Agency (USEPA) final cover regulations require a provision for the use of alternative final covers as long as they provide: 1) an infiltration layer that provides equivalent reduction in infiltration to that of the prescribed cover and 2) an erosion layer that provides equivalent protection from wind and water erosion as the one of the prescribed cover. One type of an alternative is to design landfill covers to exploit the water storage capacity of finer textured soils and the water removal capability of vegetation [2]-[6]. This type of cover is referred to as an evapotranspiration or ET cover.

Evapotranspiration (ET) covers are earthen final covers that control the percolation of infiltrating rainwater into the waste by balancing the water storage capacity of finer textured soils and the ability of plants to extract the water stored. They are designed to transmit equal or less percolation than the conventional resistive covers built with low hydraulic conductivity materials. Evapotranspiration (ET) covers (also referred to as: water balance covers, "store and release" covers, phyto-covers) are designed to have a capacity to store water by the soil and also to have the capacity to remove the stored water. Therefore, ET covers are essentially well-de- signed-vegetated soil landfill covers. The soil layer (s) of an ET cover stores moisture during rain events, and the sun and plants remove the water from the soil by evaporation and evapotranspiration, returning the moisture to the atmosphere.

Under landfill setting, final covers are also exposed to landfill gases from the waste mass below. Landfill gas is composed of 50% by volume of methane. Landfill gas is transported by diffusive and convective flux mechanisms from the anaerobic zone toward the atmosphere. The microbiological process of oxidizing CH_4 to CO_2 by methanotrophs is called microbial CH_4 oxidation. Several previous studies on landfill CH_4 oxidation in soil cover layers have demonstrated the ability of methane oxidation as a mechanism to reduce CH_4 emissions from landfill surfaces [7] -[13] . However, the capacity of landfill cover soil to oxidize CH_4 depends on both the physical and the chemical properties of landfill cover materials such as soil type, moisture content, density, organic and nutrient content, etc. Additionally, environmental conditions such as temperature and precipitation can also impact the performance of landfill cover soils in terms of oxidizing CH_4.

In ET cover context, methane oxidation is impacted by vegetation, water content and soil temperature. The soil water content and the soil temperature in the soil profile govern gas transport and therefore methane oxidation in ET covers. Water content in the ET cover soil profile is a very important factor affecting CH_4 oxidation in landfill cover soils. First, the optimum environment for CH_4-oxidizing bacteria (methanotrophic) is obtained within a specific water content range. Second, water content affects the penetration of O_2 into cover soils and O_2 is the main reactor for CH_4 oxidation. As water content increases in soil, O_2 diffusion into the soil is hindered. Third, water content affects the air-filled porosity of soil thus influences gas transport. As water fills up voided pores within the cover soil, it blocks the flow of gas upward. [14] and others correlated microbial activity with water content. Soil temperature also affects the bacterial activity of methane oxidizing bacteria. [14] and others correlated soil temperature to bacterial methane oxidation. In ET cover setting, these water content and soil temperature are dynamic and vary with time and with depth as a function of climatic conditions, vegetation, and soil hydraulic properties. In order to accurately predict methane oxidation in ET covers, landfill covers, the changing water content and temperature inside a cover has to be accounted for. In order to better predict gas emissions and methane oxidation in landfill covers, the changing water content and temperature inside the soil profile has to be accounted for in any numerical solution.

1.2. Motivations and Objectives

The potential superior capability of ET covers to reduce fugitive release of methane to the atmosphere is an additional characteristic that should be considered to encourage the use of the types of covers. The basis of ET cover design is the predictive capabilities of the water balance modeling and the ability to quantify the maximum water infiltration from the atmosphere into the waste mass below the ET cover. ET covers provide a unique opportunity to introduce a modeling approach to methane oxidation in landfill covers since the modeling approach is already undertaken during the water balance assessment of the design. The objective of this study is to use a similar modeling methodology for methane oxidation to extend the water balance modeling-based design associated with ET covers to estimate gas transport and methane oxidation in and across an ET cover. The modeling approach consists of modeling gas transport using our existing gas transport and oxidation model. Our model accounts for dynamic parameters associated with the change in water content and temperature caused by change in climatic conditions at the upper boundary of the ET cover. The model also incorporates the changing temperature and water content to methanotrophic activity in the soil. The numerical model (FSU Model) is developed by our research team at Florida State University and combines water and heat flow with a gas transport and oxidation [14] -[18] . The modeling methodology is applied to an actual ET cover test section being assessed for hydraulic equivalency in Bennington Nebraska, USA.

2. METHODS

2.1. Site Description

An ET cover was constructed and monitored as a demonstration project in Bennington, Nebraska. Bennington, NE, receives760 mm of rain per year. Snowfall is 740 mm. The number of days with any measurable precipitation is 92. On average, there are 216 sunny days per year in Bennington, NE. The July high temperature is around 30°C. The January low is −11°C. The project was intended to demonstrate that an ET cover as constructed in the demonstration project is functionally equivalent to the regulatory default cover based on water balance performance. The cover demonstration profile was constructed and instrumented in 2008. The instrumentation was activated during December 2008. The demonstration cover surface was then seeded in 2009. Surface vegetation and underlying rooting have been established. The soil moisture measurement equipment utilized on this project includes water content sensors and water potential sensors (soil suction). There were two sets of probes installed in instrumentation nests. Figure 1 shows the placement and depth spacing for the probes installed in each nest. Instrumentation nests were installed in two locations (Nest A and B). Each nest consists of 7 water content sensors and 7 water potential sensors. A water content and water potential sensor is installed at each depth indicated in Figure 1. Nest A was installed in the demonstration cover profile with a gentle slope (3.33%) that attempts to mimic the top surface of the landfill and is within that recommended by RCRA of 2 to 5%. Nest B was placed in the profile on a steeper slope (25% or 4:1) in an attempt to mimic steeper side slopes of the landfill. All of the nests of probes were connected to a central data logger. The weather station is co-located with the data logger. The equipment at this central location was mounted on a tripod in a weather resistant enclosure and lighting protection. The water balance aspect of the collected data is not addressed in this paper. Only the soil water content and potential data collected at each depth of Nest A and Nest B were averaged and used to estimate the Van Genunchten parameters describing the unstaured hydraulic properties of the ET cover layers as described later.

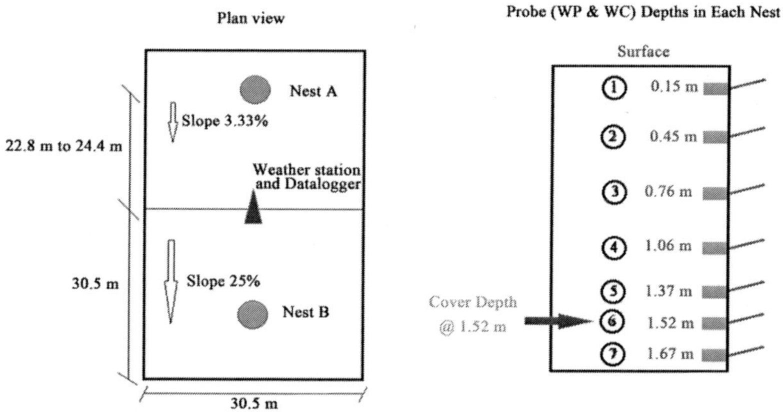

Figure 1. Plan view and profile of instrumentation nests installed in ET cover.

2.2. Model Description

A numerical model (FSU Model) was developed by our research team at Florida State University that combined water and heat flow with a gas transport and oxidation model [14] -[18] . The latest version of the FSU model was bundled via a graphical user interface and was renamed the Landfill Surface Emission Model (Land SEM).

The new Land SEM model comprises four major modules, bundled via a graphical user interface:

Ÿ A climate module which generates daily minimum and maximum air temperature along with daily rainfall based on the landfill site geographical location.

Ÿ A soil property generation module which generates soil properties, e.g. saturated hydraulic conductivity, porosity, and methane oxidation capacity based on a built-in database of soil property data from the literature. The user can also input measured soil properties, if available.

Ÿ A soil water content and temperature simulation module which uses the landfill site location, soil texture information and the climate generation data developed in the first two steps and then predicts the daily soil moisture and temperature at any depth of the soil cover profile for an average climatic year.

Ÿ A core computational module, based on [14] -[18] , which simulates the concentration and flux of the gas components including methane in the landfill cover soil profile.

The climate simulator is a module adopted from the models Global Temp SIM and Global Rain [19] , including an air temperature simulation and precipitation simulation. The air temperature generation predicts the daily average air temperature based on a 30-year (1961-1990) temperature records and calculate the daily maximum and minimum air temperature. The global rainfall module estimates daily rainfall patterns for a yearly cycle based on a 30-year (1961-1990) global average monthly precipitation records. The information obtained from these modules together with the soil properties are used in the soil water content and temperature simulators. Global Temp SIM and Global Rain were used to estimate daily average air temperatures and average daily precipitation for Bennington, Nebraskausing the climate generation module with the USA postal service Zip Code as the main input.

The continuity equation describing the reactive transport of gas component i in landfill cover soils can be written [14]

$$\varepsilon \frac{\partial C_i}{\partial t} = -\nabla \cdot J_i + \rho_{\text{bulk}} \cdot r_i \tag{1}$$

where ε is air-filled porosity $\left(\text{m}^3_{\text{gas}}/\text{m}^3_{\text{soil}}\right)$, C_i is the molar concentration of gas component i $\left(\text{mole}\cdot\text{m}^{-3}\right)$, J_i is the flux of gas component i due to advection and diffusion $\left(\text{mol}\cdot\text{m}^{-2}\cdot\text{s}^{-1}\right)$, ρ_{bulk} is the bulk density of soil $\left(\text{kg}\cdot\text{m}^{-3}\right)$, and r_i is the production rate of gas component I $\left(\text{mol}\cdot\text{kg}^{-1}_{\text{drys}}\cdot\text{s}^{-1}\right)$. In Equation (1), the first

time on right hand side represents the transport process while the second term accounts for the soil gas reaction (oxidation for methane).

The flux J_i can be determined by

$$J_i = -D_{\mathrm{soil},i}\nabla C_i + vC_i \tag{2}$$

where $D_{\mathrm{soil},i}$ is the diffusion coefficient of component i in soil $\left(\mathrm{m^2 \cdot s^{-1}}\right)$, v is the flow velocity of gas mixture in the soil (m/s).

The diffusion coefficient $D_{\mathrm{soil},i}$ can be calculated from its relationship with the diffusion coefficient in gas mixture,

$$D_{\mathrm{soil},i} = \gamma D_{\mathrm{gas},i} \tag{3}$$

where γ is the relative diffusion coefficient of gas in a tortuous void space in soil and $D_{\mathrm{gas},i}$ is the diffusion coefficient of gas component i in gas mixture. The relative diffusion coefficient γ has been found to be independent to the gas itself [20] [21] . Although several formulation of γ based on air content in soils have been proposed (e.g. [22] [23]), the following equations by [24] is used in this research,

$$\gamma = \frac{\varepsilon^{2.5}}{\phi} \tag{4}$$

where ϕ is total porosity $\left(\mathrm{m^3_{void}/m^3_{soil}}\right)$.

The gas flow velocity is due to advection and is assumed to follow Darcy's law.

$$v = -\frac{k}{\mu}\frac{\partial P}{\partial x} \tag{5}$$

where k is the intrinsic permeability of soil (m^2), μ is the gas-mixture viscosity, P is pressure (Pa) which is calculated based on the ideal gas law

$$P = \sum_{i=1}^{n} C_i \cdot R \cdot T \tag{6}$$

where R is the universal gas constant (8.314 $\mathrm{J \cdot K^{-1} \cdot mol^{-1}}$) and T is absolute temperature (K).

The reaction rate of CH_4 is given

$$r_{CH_4} = V_{max} \frac{C_{CH_4}}{K_{m[CH_4]} + C_{CH_4}} \frac{C_{O_2}}{K_{m[O_2]} + C_{O_2}} \quad (7)$$

where V_{max} is the maximum methane consumption rate $(nmol \cdot s^{-1} \cdot g^{-1}_{dry\ soil})$; and $K_{m[CH_4]}$ and $K_{m[O_2]}$ are the half-saturation constants of CH_4 and O_2 $(mol \cdot m^{-3})$, respectively. Methane oxidation is affected by many factors among which soil temperature and moisture content are more significant. Temperature, moisture, and scaling correction factors were considered for the kinetic parameters in Equation (7) as described in [18] . [14] developed a mechanism to relate in situ rates of oxidation to those measured in laboratory incubation experiments on homogenized and sieved soil samples under fixed environmental conditions. Their study used combined laboratory and field measurements to develop field scale Michaelis-Menten parameters to be used in conjunction with numerical models to predict methane emissions and oxidation in landfill covers and under field conditions. More details about the model can be found in [14] -[18] . Since the surface of the phytocap is open to the atmosphere, the gas composition above the surface node of the model was assumed to be the atmospheric gas compositions (background), which are 21.2%v/v oxygen, 1.8 ppmv methane, 385.0 ppmv carbon dioxide and 78.8%v/v nitrogen.

2.3. Simulations

The simulated ET cover was divided into five30-cm-thick layers with different soil properties. The top layer was assumed to be 30 cm thick and represented by the 15 cm deep soil water and soil suction sensors. The second layer was also assumed to be 30 cm thick and be represented by the sensors buried at 45 cm, etc. The top 60 cm ET cover soil was assumed to have undergone several freeze-thaw and desiccation cycles. The long-term saturated hydraulic conductivity of the top 30-cm-thick layer was assigned to be 1×10^{-4} cm/s. The second layer was a signed to have a saturated hydraulic conductivity of 1×10^{-5} cm/s. The third layer was assumed to have a 10 times lower saturated hydraulic conductivity. The hydraulic conductivity of the lowest layer was measured from an undisturbed sample as 2.7×10^{-7} cm/s. The hydraulic conductivity of the second layer from the bottom was assumed to be 5 times higher than the bottom layer. The porosity of the soil profile was assumed as follow: from top to down 0.45, 0.42, 0.40, 0.38, and 0.38. The long term unsaturated soil properties are shown in Figure 2. The collected daily moisture content (moisture content sensor) and soil suction (matric potential sensor) value for a given elevation (each probe depth) were used to directly measure a field determined soil water characteristic curve (Figure 2). The in situ determination of a SWCC for the soil is more representative of the actual conditions than the

SWCC developed in the laboratory. Figure 2 shows the soil properties used as input into the model for all the simulated layers of the ET cover and three examples of field determined SWCCs.

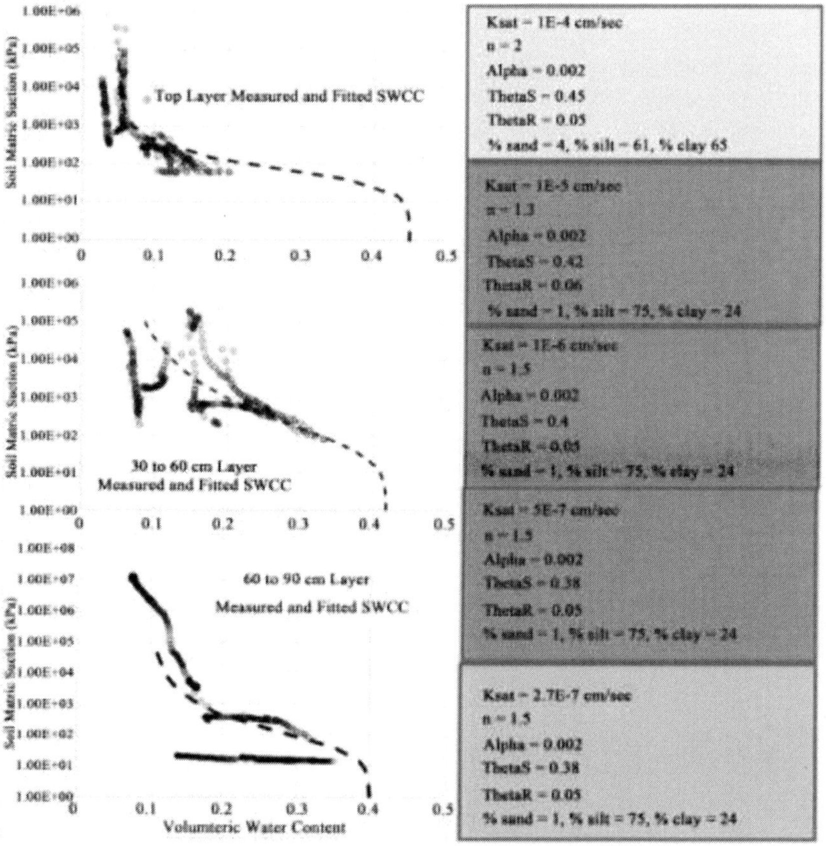

Figure 2. Examples of measured and fitted SWCC obtained using collected data, along with all soil properties used for each layer of the simulated ET cover.

One of the major inputs to the gas transport model is the maximum methane oxidation capacity (a lab measured quantity V_{max}) of the soil. This quantity represents the maximum capacity of bacteria in the soil to oxidize methane at room temperature. Based on our experience with measuring V_{max} on landfill cover soils and for the purpose of this document, different simulations of two different values of V_{max} (150, 300 nmol·s^{-1}·g^{-1} dry soil) were performed. Several simulations were performed using different rates of methane loading into the ET cover. For each V_{max}, several simulations were performed using methane loading varying from 2 to 200 $gm^{-2} \cdot d^{-1}$. The landfill gas entering the bottom of soil profile was partitioned equally to CH_4 and CO_2 v/v.

3. RESULTS AND ANALYSIS

The output of the gas transport model consists of daily surface emissions of methane and daily percent oxidized for each simulation performed with each methane loading flux (MLF) during the modelled average one year period in Bennignton, NE. The daily simulations was repeated for different MLFs (2, 5, 10, 20, 50, 100, 200 $gm^{-2} \cdot d^{-1}$) and with each V_{max}. Figure 3 and Figure 4 show the daily surface emissions and daily percent methane oxidation for all the simulations with different MLFs and with V_{max} of 150and 300 $nmol \cdot s^{-1} \cdot g^{-1}$ dry soil, respectively.

Figure 3 and Figure 4 show that for MLF values of 2 to 5 $gm^{-2} \cdot d^{-1}$, noneto very little surface emissions occur during the entire year. The ET cover seems to be able to oxidize all of the incoming methane from below each day of the year, even during cold seasons and inside and outside the growing season. Complete oxidation (100%) occurs every day of the year when the MLF is lower than around 5 $gm^{-2} \cdot d^{-1}$. For simulations with a higher MLF values, the highest daily surface emissions and lower percent oxidation occur outside the growing season and during colder periods of the year. The higher percent oxidation and the lowest surface emissions occur during the growing season and during the spring and summer seasons indicating the effects of soil temperature and water contents on surface emissions and therefore methane oxidation.

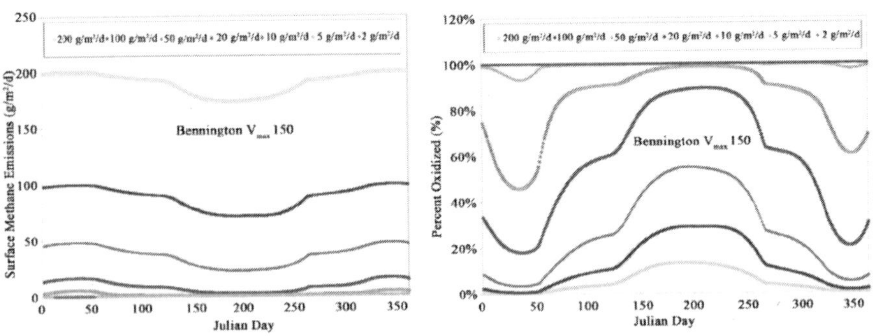

Figure 3. Surface emissions, percent oxidized for several methane loading fluxes with V_{max} = 150 $nmol \cdot s^{-1} \cdot g^{-1}$ dry soil.

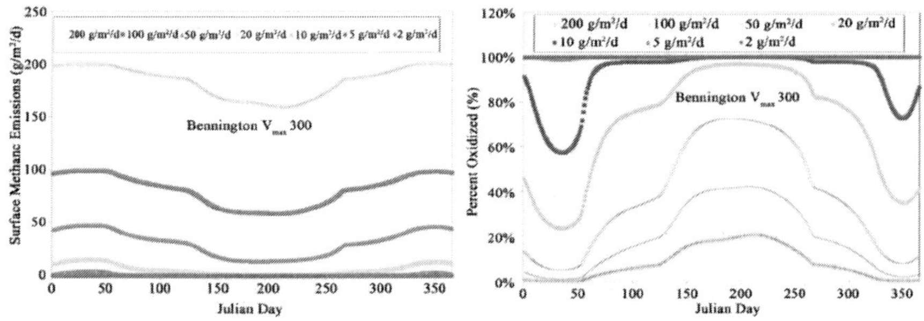

Figure 4. Surface emissions, percent oxidized for several methane loading fluxes with V_{max} = 300 nmol·s^{-1}·g^{-1} soil.

The data presented in Figure 3 and Figure 4 were then used to calculate average yearly methane oxidation rate and average yearly percent oxidation for each simulated methane loading. The average yearly methane oxidation rate in gm^{-2}d^{-1}was determined for each MLF and plotted in Figure 5. Figure 5 shows that at lower MLFs, oxidation rate is equal to the MLF and increases with increasing MLF. In this region, the ET cover is able to oxidize all of the methane coming from below. In this region, the slope of the curve shown in Figure 5 is 1:1. In this region of low MLFs, there is no difference between the simulations with 150 and 300 nmol·s^{-1}·g^{-1} dry soilsince both ET covers can oxidize all of the incoming methane. There seems to be an MLF below which methane oxidation is total (around 7 - 8 gm^{-2}·d^{-1} for both V_{max} values). The methane loading flux at which the cover can oxidize all of the incoming methane can be considered as the MLF below which no surface emissions are measurable (No surface emissions). Above this MLF, the ET cover is not able to oxidize all of the incoming methane but still can oxidize more and more methane as the MLF increases. Figure 5 also shows that each ET cover soil seems to have a maximum capacity to oxidize the incoming methane depending on V_{max}. Figure 5 shows that for each V_{max}, there is a point where the ET cover will not oxidize any more methane, which indicates a certain capacity of the cover soil (around 13 gm^{-2}·d^{-1} and 19 gm^{-2}·d^{-1} for V_{max} of 150 and 300 respectively). Figure 5 also shows that oxidation rate starts to decrease after a certain MLF. We believe that this is due to the fact that these results are obtained using a pressure based model. The pressure below the landfill cover is increased to increase the MLF into the cover soil. At a certain pressure (depending on the cover) the incoming flux of methane starts to inhibit the diffusion of oxygen from the upper boundary and limit the amount of methane that can be oxidized.

The shape of the graph in Figure 5 resembles a Michaelis-Menten enzyme kinetic plot of a biological reactions response to increasing concentrations of a substrate. In landfill ET cover, soil methane oxidation is conducted by a biological system and it behaves as such. When substrate (CH_4) is supplied, the cover's rate of CH_4 uptake is linear to a point, and then the system apparently becomes saturated. This response must be due to the oxygen levels to which the bacteria are exposed in addition to temporal variability in soil temperature and moisture.

The same simulation results were used to calculate the average percent year methane oxidation, and plotted versus MLF for each V_{max} in Figure 6. Figure 6 shows that there seems to be a methane loading rate below which methane oxidation is equal to 100%. Above this methane loading, the cover is not able to oxidize all of the incoming methane. Therefore, percent oxidation starts to decrease as the methane loading increases. As expected, at any given MLF, ET covers with higher V_{max} have higher percent methane oxidation.

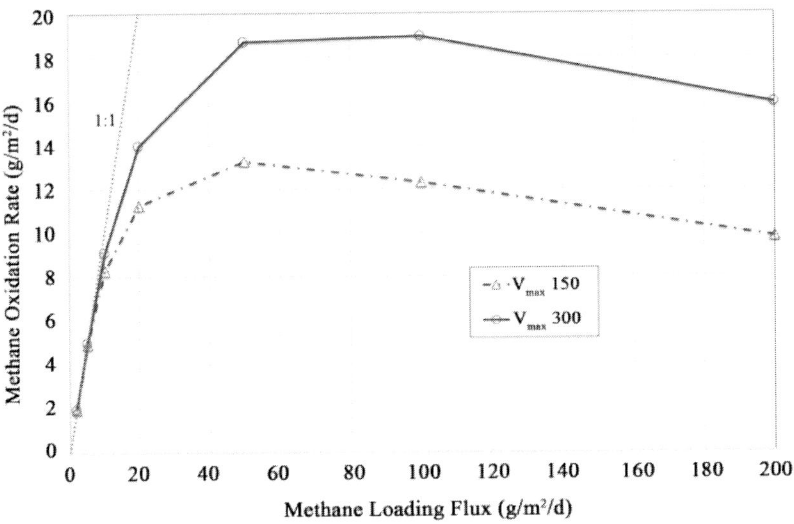

Figure 5. Methane oxidation rate versus MLF for simulations with V_{max} = 150 and 300 nmol·s^{-1}·g^{-1} dry soil.

4. CONCLUSIONS AND PRACTICAL IMPLICATIONS

One way to estimate uncontrolled emissions or fugitive emissions from the surface of landfills is to use the AP- 42 protocol proposed by the United States Environmental Protection Agency (USEPA, 1997). AP-42, when applied to landfill methane surface emissions, can be written as follows:

$$CM_P = \left[UM_P * \left(1 - \frac{\eta_{col}}{100}\right)\right] * (1 - f_{ox}) \qquad (8)$$

where CM_P is the controlled mass emissions of methane (kg/year), UM_P is the methane generation from the waste mass in kg/year (estimated by LANDGEM

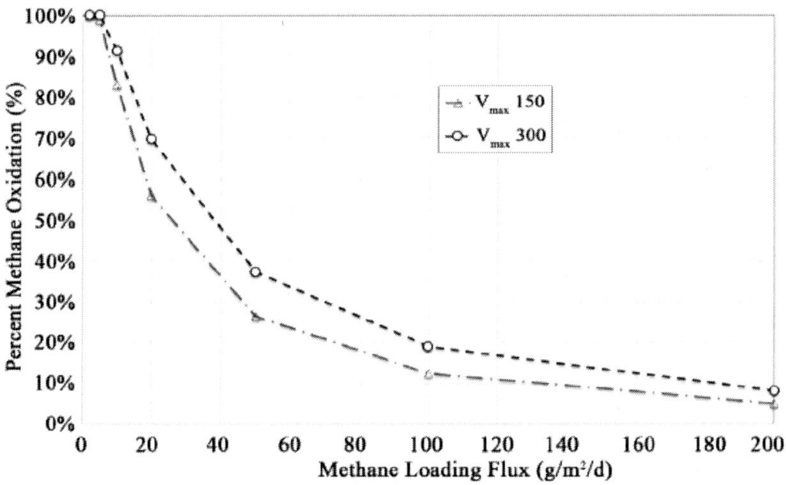

Figure 6. Percent methane oxidation versus MLF for simulations with V_{max} = 150 and 300 nmol·s^{-1}·g^{-1} dry soil.

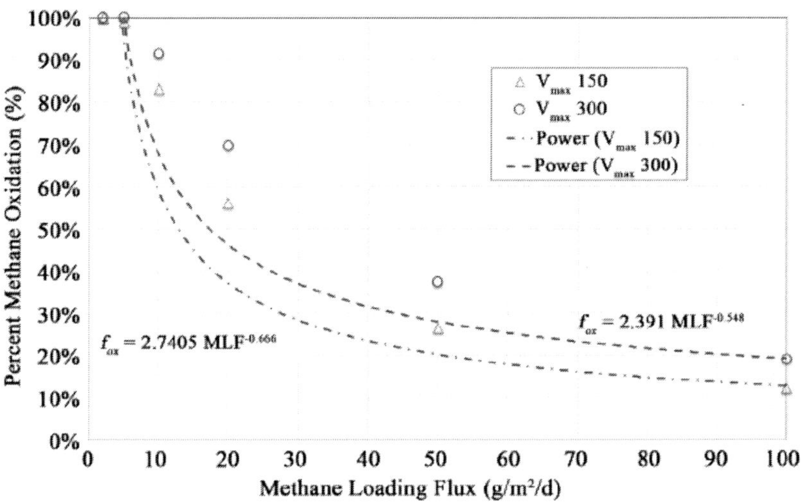

Figure 7. Correlation of Percent methane oxidation versus MLF for simulations with V_{max} = 150 and 300 nmol·s^{-1}·g^{-1} dry soil.

or any other allowed model), η_{col} is the collection efficiency of the landfill gas collection system. The collection efficiency of the gas collection system can be assumed the default values of 75% assigned for soil covers. The remaining 25%

of the generated methane can be assumed to escape through the landfill ET cover such as the one simulated in this study, f_{ox} is the percent oxidation or fraction of methane oxidized in the cover soil. The USEPA has also allowed the use of 10% to account for methane oxidation that can occur in landfill cover soils.

One application of this model is to be used in conjunction of AP-42 guidelines but with a modeled methane oxidation, f_{ox} instead of the default value of 10% typically assumed in AP-42. The use of the correlations of f_{ox}, shown in Figure 7, can encourage the use of ET covers and provide an estimate of fugitive methane emissions, along with quantifying the improvement of performance (emissions and oxidation) of ET covers. Such protocol can help decision makers quantify additional methane oxidation capabilities of ET covers. The modeled percent oxidation for each V_{max} was expressed as a power function of methane loading flux (MLF) as shown in Figure 7. The methane loading flux is estimated as the 25% of generation (methane escaping the gas collection system) divided by the surface area of the ET covered landfill.

REFERENCES

1. 1. Albrecht, B. and Benson, C. (2001) Effect of Desiccation on Compacted Natural Clays. J. of Geotech. and Geoenv. Eng., ASCE, 127, 67-76.
2. Benson, C. and Khire, M. (1995) Earthen Covers for Semi-Arid and Arid Climates, ASCE 1995 National Convention on Landfill Closures-Environmental Protection and Land Recovery. Geotechnical Special Publication No. 53, 201-217.
3. Stormont, J. and Morris, C. (1998) Method to Estimate Water Storage Capacity of Capillary Barriers. Journal of Geotechnical and Geoenvironmental Engineering, 124, 297-302.
4. Nyhan, J.W., Schofield, T.G. and Starmer, R.H. (1997) A Water Balance Study of Four Landfill Cover Designs Varying in Slope for Semi-Arid Regions. Journal of Environmental Quality, 26, 1385-1392.
5. Ward, A.L. and Gee, G. (1997) Performance Evaluation of a Field-Scale Surface Barrier. Journal of Environmental Quality, 26, 694-705.
6. Hauser, V., Weand, B., Shaw, M. and Wusterbarth, A. (1996) Natural Covers for Landfills: A Closer Look. Proceedings of the 22nd Symposium and Exhibition, American Defense Preparedness Association, Orlando, March 1996.

7. Bogner, J., Spokas, K., Burton, E., Sweeney, R. and Corona, V. (1995) Landfills as Atmospheric Methane Sources and Sinks. Chemosphere, 31, 4119-4130.
8. Borjesson, G. and Svensson, B. (1997) Seasonal and Diurnal Methane Emissions from a Landfill and Their Regulation by Methane Oxidation. Waste Management and Research, 15, 33-54.
9. Abichou, T., Chanton, J., Powelson, D., Fleiger, J., Escoriaza, S., Lei, Y. and Stern, J. (2006) Methane Flux and Oxidation at Two Types of Intermediate Landfill Covers. Waste Management, 26, 1305-1312.
10. Kjeldsen, P., Dalager, A. and Broholm, K. (1997) Attenuation of Methane and Nonmethane Organic Compounds in Landfill Gas Affected Soils. Journal of Air & Waste Management Association, 47, 1268-1275
11. Scheutz, K., Kjeldsen, P., Chanton, J., Blake, D. and Bogner, J. (2003) Comparative Oxidation and Net Emissions of CH4 and Selected Non-CH4 Organic Compounds in Landfill Cover Soils. Environmental Science and Technology, 37, 5150-5158.
12. Huber-Humer, M., Gebert, J. and Hilger, H. (2008) Biotic Systems to Mitigate Landfill Methane Emissions. Water Management Research, 26, 33-46.
13. 13. Bogner, J. and Spokas, K. (1993) Landfill CH4: Rates, Fates, and Role in Global Carbon Cycle. Chemosphere, 26, 369-386.
14. Abichou, T., Mahieu, K., Chanton, J., Romdhane, M. and Mansouri, I. (2011) Scaling Methane Oxidation: From Laboratory Incubation Experiments to Landfill Cover Field Conditions. Waste Management, 31, 978-986.
15. 15. Abichou, T., Mahieu, K., Yuan, L., Chanton, J. and Hater, G. (2009) Effects of Compost Biocovers on Gas Flow and Methane Oxidation in a Landfill Cover. Waste Management, 29, 1595-1601.
16. Yuan, L., Abichou, T., Chanton, J., Powelson, D. and De Visscher, A. (2009) Long Term Numerical Solution of Methane Transport and Oxidation in a Compost Biofilter. Practice Periodical of Hazardous Toxic and Radioactive Waste Management, 13, 196-202.
17. Abichou, T., Yuan, L. and Chanton, J. (2008) Estimating Methane Emission and Oxidation from Earthen Landfill Covers. Geocongress 2008: Geotechnics of Waste Management and Remediation (GSP), New Orleans, 9-12 March 2008, 80-87.

18. Abichou, T., Johnson, T., Mathieu, K., Chanton, J., Romdhane, M. and Mansouri, I. (2010) Developing a Design Approach to Reduce Methane Emissions from California Landfills. In: Fratta, D., Muhuntan, B. and Pupala, A., Eds., GeoFlorida 2010: Advances in Analysis, Modeling & Design, ASCE, Geotechnical Special Publication No. 199, Baltimore, 2878-2887.

19. 19. Spokas, K. and Forcella, F. (2009) Software Tools for Weed Seed Germination Modeling. Weed Science, 57, 216-227.

20. Jin, Y. and Jury, A. (1996) Characterizing the Dependence of Gas Diffusion Coefficient on Soil Properties. Soil Science Society of America Journal, 60, 66-71.

21. Stein, V. B., Hettiaratchi, J. and Achari, G. (2003) Numerical Model for Biological Oxidation and Migration of Methane in Soils. Practice Periodical of Hazardous, Toxic, and Radioactive Waste Management, 5, 225-234.

22. Freijer, J. (1994) Calibration of Jointed Tube Model for the Gas Diffusion Coefficient in Soils. Journal of the American Society of Soil Science, 58, 1067-1076.

23. Kightley, D., Nedwell, D. and Cooper, M. (1995) Capacity for Methane Oxidation in Landfill Cover Soils Measured in Laboratory Scale Soil Microcosms. Journal of Applied Environmental Microbiology, 61, 592-601.

24. Moldrup, P., Olesen, T., Gamst, J., Schjønning, P., Yamaguchi, T. and Rolston, D.E. (2000) Predicting the Gas Diffusion Coefficient Inrepacked Soil: Water-Induced Linear Reduction Model. Soil Science Society of America Journal, 64, 1588-1594.

Chapter 11

A Coupled Remote Sensing and Simplified Surface Energy Balance Approach to Estimate Actual Evapotranspiration from Irrigated Fields

Gabriel B. Senay [1,*], Michael Budde [2], James P. Verdin [3] and Assefa M. Melesse [4]

[1]SAIC, contractor to U.S. Geological Survey (USGS) Center for Earth Resources Observation and Science (EROS)/Geographic Information Science Center of Excellence (GIScCE), South Dakota State University, Sioux Falls, SD, USA. Work performed under USGS contract 03CRCN0001; Tel: (605) 594-2758

[2]SAIC/U.S. Geological Survey EROS, Sioux Falls, SD, USA; Tel: (605) 594-2619

[3]U.S. Geological Survey EROS, Sioux Falls, SD, USA; Tel: (605) 594-6018

[4]Department of Environmental Studies, Florida International University, Miami, FL, USA; Tel: (305) 348-6518

ABSTRACT

Accurate crop performance monitoring and production estimation are critical for timely assessment of the food balance of several countries in the world. Since 2001, the Famine Early Warning Systems Network (FEWS NET) has been monitoring crop performance and relative production using satellite-derived data and simulation models in Africa, Central America, and Afghanistan where ground-based monitoring is limited because of a scarcity of weather stations. The commonly used crop monitoring models are based on a crop water-balance algorithm with inputs from satellite-derived rainfall estimates. These models are useful to monitor rainfed agriculture, but they are ineffective for irrigated areas. This study focused on Afghanistan, where over 80 percent of agricultural production comes from irrigated lands. We developed and implemented a Simplified Surface Energy Balance (SSEB) model to monitor and assess the performance of irrigated agriculture in Afghanistan using a combination of 1-km thermal data and 250-m Normalized Difference Vegetation Index (NDVI) data, both from the Moderate Resolution Imaging Spectroradiometer (MODIS) sensor. We estimated seasonal actual evapotranspiration (ETa) over a period of six years (2000-2005) for two major irrigated river basins in Afghanistan, the Kabul and the Helmand, by analyzing up to 19 cloud-free thermal and NDVI images from each year. These seasonal ETa estimates were used as relative indicators of year-to-year production magnitude differences. The temporal water-use pattern of the two irrigated basins was indicative of the cropping patterns specific to each region. Our results were comparable to field reports and to estimates based on watershed-wide crop water-balance model results. For example, both methods found that the 2003 seasonal ETa was the highest of all six years. The method also captured water management scenarios where a unique year-to-year variability was identified in addition to water-use differences between upstream and downstream basins. A major advantage of the energy-balance approach is that it can be used to quantify spatial extent of irrigated fields and their water-use dynamics without reference to source of water as opposed to a water-balance model which requires knowledge of both the magnitude and temporal distribution of rainfall and irrigation applied to fields.

Keywords: evapotranspiration; energy balance; SSEB; reference ET; irrigation

1. INTRODUCTION

As reconstruction efforts in Afghanistan continue, the need for an accounting of the country's water balance is necessary to address current and future water use scenarios. One aspect of the water balance equation involves an assessment of the water demands by irrigated agricultural lands. Since large and widely dispersed populations depend on rainfed and irrigated agriculture and pastoralism, large-area monitoring and forecasting are important inputs to such assessments. The Famine Early Warning Systems Network (FEWS NET), an activity funded by the United States Agency for International Development (USAID), employs a crop water-balance model (based on the water demand and supply at a given location) to monitor the performance of rainfed agriculture and forecast relative production before the end of the crop growing season. Although a crop water-balance approach is effective for monitoring rainfed agriculture [1, 2], irrigated agriculture is best monitored by other methods since the supply (water used for irrigation) is usually generated from upstream areas, which are distant from the demand location.

The surface energy balance method has been successfully applied by several researchers [3-6] to estimate crop water use in irrigated areas. Their approach requires solving the energy balance equation at the surface (Equation 1) where the actual evapotranspiration (ETa) is calculated as the residual of the difference between the net radiation to the surface and losses due to the sensible heat flux (energy used to heat the air) and ground heat flux (energy stored in the soil and vegetation).

$$LE = Rn - G - H \tag{1}$$

- LE = Latent heat flux (energy consumed by evapotranspiration) (W/m^2)
- Rn = Net radiation at the surface (W/m^2)
- G = Ground heat flux (W/ m^2)
- H = Sensible heat flux (W/ m^2)

The estimation of each of these terms from remotely sensed imagery requires high quality data sets. Allen et al. [3] described the several steps required to estimate actual ET using the surface energy-balance method that employs the hot and cold pixel approach of Bastiaanssen et al. [4]. In summary, for the net radiation term, data on incoming and outgoing radiation and the associated surface albedo and emissivity fractions for shortwave and long wave bands are required. The ground heat flux is estimated using surface temperature, albedo, and normalized difference vegetation index (NDVI). The sensible heat flux is estimated as a function of the temperature gradient above the surface, surface roughness, and wind speed.

Although solving the full energy-balance approach has been shown to give good results in many parts of the world, the data and skill requirements to solve for the various terms in the equation are prohibitive for operational applications where year-to-year differences and long term anomalies are more useful than absolute values. In this study, we developed and implemented a simplified version of the surface energy-balance approach to estimate actual ET while maintaining and extending the major assumptions in the Surface Energy Balance Algorithm for Land (SEBAL, Bastiaanssen et al. [4]) and the Mapping Evapotranspiration at High Resolution using Internalized Calibration (METRIC, Allen et al. [3]) method. Both methods assume that the temperature difference between the land surface and the air (near-surface temperature difference) varies linearly with land surface temperature. They derive this relationship based on two anchor pixels known as the hot and cold pixels, representing dry and bare agricultural fields and wet and well-vegetated fields, respectively. The SEBAL and METRIC methods use the linear relationship between the near-surface temperature difference and the land surface temperature to estimate the sensible heat flux which varies as a function of the near-surface temperature difference, by assuming that the hot pixel experiences no latent heat, i.e., ET = 0.0, whereas the cold pixel achieves maximum ET. Similarly, Su et al. [7] have developed a Surface Energy Balance System (SEBS) that refers to the hot and cold extreme conditions as "dry" and "wet", respectively.

In this study, we extended this assumption with a simplification by stating that the latent heat flux (actual evapotranspiration) also varies linearly between the hot and cold pixels. This assumption is based on the logic that the temperature difference between soil surface and air is linearly related to soil moisture [8]. Furthermore, crop soil water-balance methods estimate actual ET using a linear reduction from the potential ET depending on soil moisture [12, 1]. Therefore, we argue that actual ET can be estimated by the near-surface temperature difference, which in turn is estimated from the land surface temperatures of the hot and cold pixels in the study area. In other words, while the hot pixel of a bare agricultural area experiences little ET and the cold pixel of a well-watered irrigated field experiences maximum ET, and the remaining pixels in the study area will experience ET in proportion to their land surface temperature in relation to the hot and cold pixels. This approach can be compared to the crop water stress index (CWSI) first developed by Jackson [9]. The CWSI is derived from the temperature difference between the crop canopy and the air. Dividing the current temperature difference (canopy vs air) by known upper and lower canopy air temperature difference values creates a ratio index varying between 0 and 1. The lower limiting canopy temperature is reached when the crop transpires without water shortage, and the upper limiting canopy temperature is reached when the plant transpiration is zero, owing to water shortage, Qiu et al. [10]. In this study, the cold and hot anchor land surface temperature pixel values are the equivalent of the lower and upper limiting canopy temperatures of the CWSI method.

The main objective of this study was to produce actual evapotranspiration estimates using a combination of a simplified surface energy balance (SSEB) approach and remotely-sensed MODIS thermal imagery and global reference ET over known irrigated fields in Afghanistan.

2. METHODS

2.1. Study sites

The study focuses on two major expanses of irrigation in the Helmand and Kabul basins (Figure 1). As mentioned, the water use by irrigated agricultural lands is one aspect of the total water balance in these areas. This work was done as part of a larger Afghanistan water project that identified these two basins as those of primary concern. Because this method relies heavily on

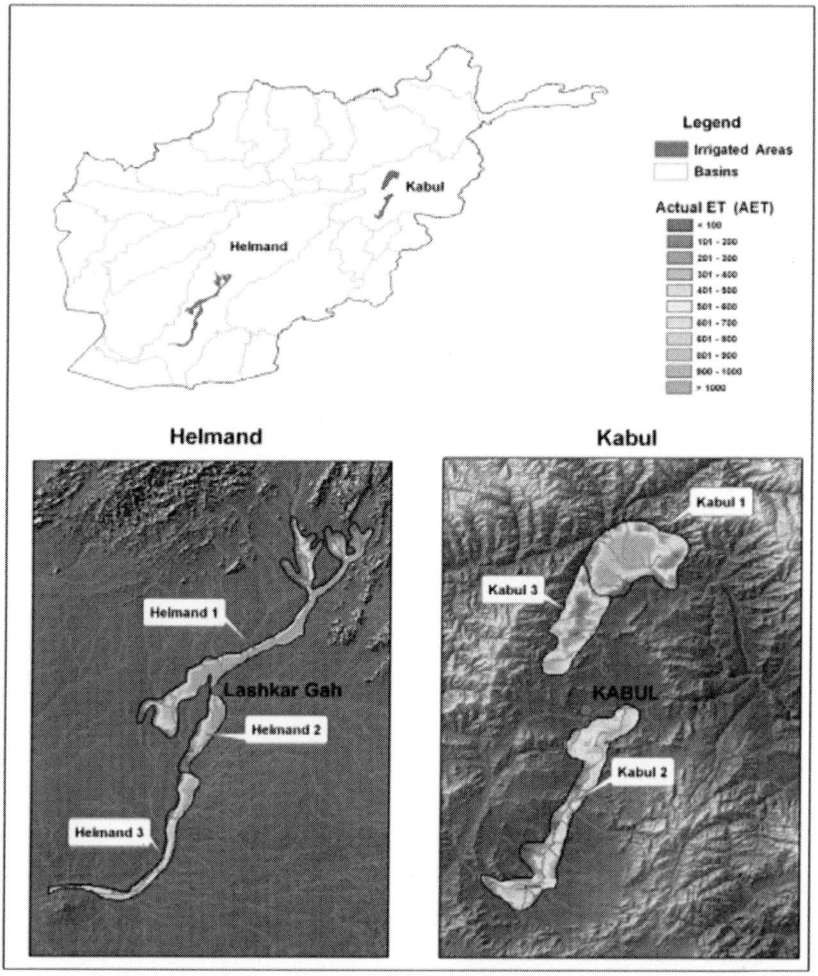

Figure 1. Study sites showing the irrigated fields in the Helmand and Kabul drainage basins. Enlarged windows illustrate the three sub-divisions within each irrigated area and their relation to stream networks.

temperature variability, each of these areas was further sub-divided into 3 sections of irrigated cropland in order to minimize the effects of elevation differences on surface temperature measures. Table 1 illustrates differences in area and median elevation for each of the sub-divisions. For each sub-division, a polygon was defined around the irrigated fields using a combination of Landsat, Advanced Spaceborne Thermal Emission and Reflection Radiometer (ASTER), and Moderate Resolution Imaging Spectroradiometer (MODIS) data sets. The irrigated areas consist of both well-vegetated and sparsely vegetated areas, with some arid/semi-arid areas at the periphery. In the case of the

Helmand basin, Helmand 1 is comprised of the area upstream of Lashkar Gah, Helmand 2 is an approximately 35km stretch of irrigated land south of Lashkar Gah along the main river channel, and Helmand 3 is the remainder of the irrigated area along the channel extending to the south and west. The Kabul basin sub-divisions consist of an extensive irrigated area north of Kabul and a narrower band of irrigation along the river south and west of Kabul. The latter area, Kabul 2, is adjacent to one of the primary stream channels feeding the city and could have significant impacts on municipal water supply. Kabul 1 and Kabul 3 include the large area of irrigated agriculture that is found near the headwaters of the Panjir River. Wheat is the primary cultivated crop in all areas.

Table 1. Study area characteristics by irrigation area sub-division.

Sub-Division	Area (sq km)	Median Elevation (m)
Kabul 1	741	1508
Kabul 2	864	1896
Kabul 3	448	1671
Helmand 1	2120	860
Helmand 2	429	746
Helmand 3	719	673

2.2. Data Set Characteristics

The primary data sets for this study were derived from the MODIS sensor flown onboard the Terra satellite. MODIS Land Surface Temperature (LST) data were used to calculate the crucial evapotranspiration (ET) fractions explained in the procedures section of this manuscript. Additionally, MODIS NDVI data were used for irrigated area delineation and identifying highly-vegetated versus sparsely vegetated areas within the agricultural zone. The global reference ET data were obtained from the archives of USGS/FEWS NET operational model outputs. Each data set is further described below.

MODIS Land Surface Temperature: Thermal surface measurements were collected from the MODIS 8-day Land Surface Temperature/Emissivity (LST/E) product (MOD11A2). The MODIS instrument provides 36 spectral bands, including 16 in the thermal portion of the spectrum. The LST/E images provide per-pixel temperature and emissivity values at 1-km spatial resolution for the 8-day composite product. Temperatures are extracted in degrees Kelvin with a view-angle dependent algorithm applied to direct observations. This study utilized average daytime land surface temperature measurements for 8-day composite periods throughout the growing season (Figure 2). More than twenty five 8-day periods beginning in early April through the end of October were processed. Table 2 shows the MODIS day of year representing the first day of the 8-day MODIS LST composite period and corresponding calendar dates for all years.

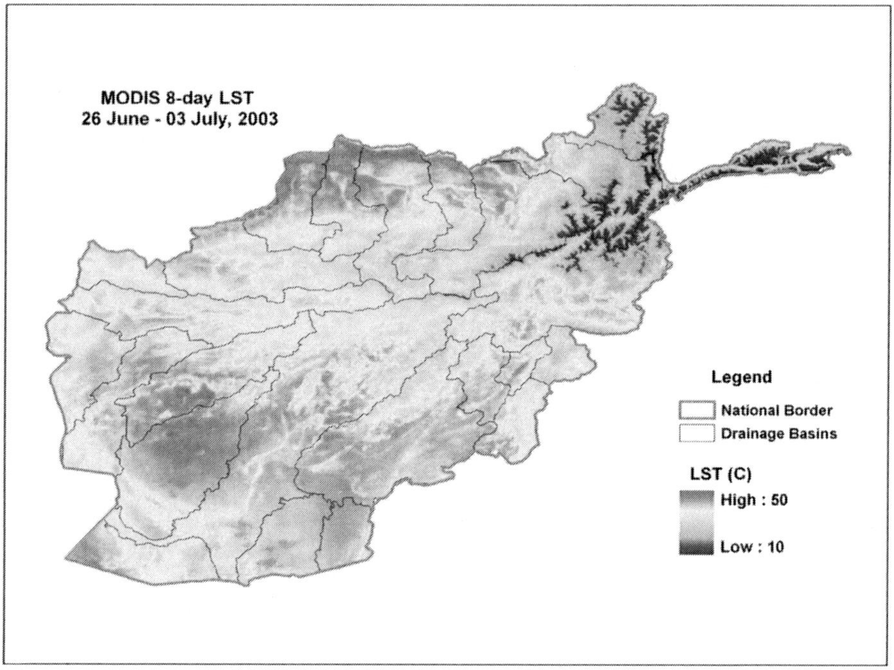

Figure 2. An 8-day composite of MODIS Land Surface Temperature (LST, °C) in late June/early July 2003.

MODIS Vegetation Index: MODIS Vegetation Index (VI) products use reflectance measures in the red (620 – 670 nm), near infrared (841 – 876 nm), and blue (459 – 479 nm) bands to provide spectral measures of vegetation vigor. The MODIS VI products include the standard normalized difference vegetation index (NDVI) and the enhanced vegetation index (EVI). Both indices are available at 250-m, 500-m, and 1-km spatial resolution. The primary difference between the two indices is that EVI uses blue reflectance to provide better sensitivity in high biomass regions. Because this study was concentrated on irrigated agriculture in an otherwise dry land environment, we used the standard 16-day NDVI product at 250-m resolution for this analysis (Figure 3).

Table 2. MODIS composite day and corresponding calendar dates.

MODIS Composite Day	2001, 2002, 2003, 2005	2000, 2004
89	Mar 30 - Apr 06	Mar 29 - Apr 05
97	Apr 07 - 14	Apr 06 - 13
105	Apr 15 - 22	Apr 14 - 21
113	Apr 23 - 30	Apr 22 - 29
121	May 01 - 08	Apr 30 - May 07
129	May 09 - 16	May 08 - 15
137	May 17 - 24	May 16 - 23
145	May 25- Jun 01	May 24 - 31
153	Jun 02 – 09	Jun 1 - 08
161	Jun 10 – 17	Jun 09 - 16
169	Jun 18 – 25	Jun 17 - 24
177	Jun 26 - Jul 03	Jun 25 - Jul 02
185	Jul 04 - 11	Jul 03 - 10
193	Jul 12 - 19	Jul 11 - 18
201	Jul 20 - 27	Jul 19 - 26
209	Jul 28 - Aug 04	Jul 27 - Aug 03
217	Aug 05 - 12	Aug 04 - 11
225	Aug 13 - 20	Aug 12 - 19
233	Aug 21 - 28	Aug 20 - 27
241	Aug 29 - Sep 05	Aug 28 - Sep 04
249	Sep 6 - 13	Sep 05 - 12
257	Sep 14 - 21	Sep 13 - 20
265	Sep 22 - 29	Sep 21 - 28
273	Sep 30 - Oct 07	Sep 29 – Oct 06
281	Oct 08 - 15	Oct 07 – 14
289	Oct 16 - 23	Oct 15 – 22
297	Oct 24 - 31	Oct 23 – 30

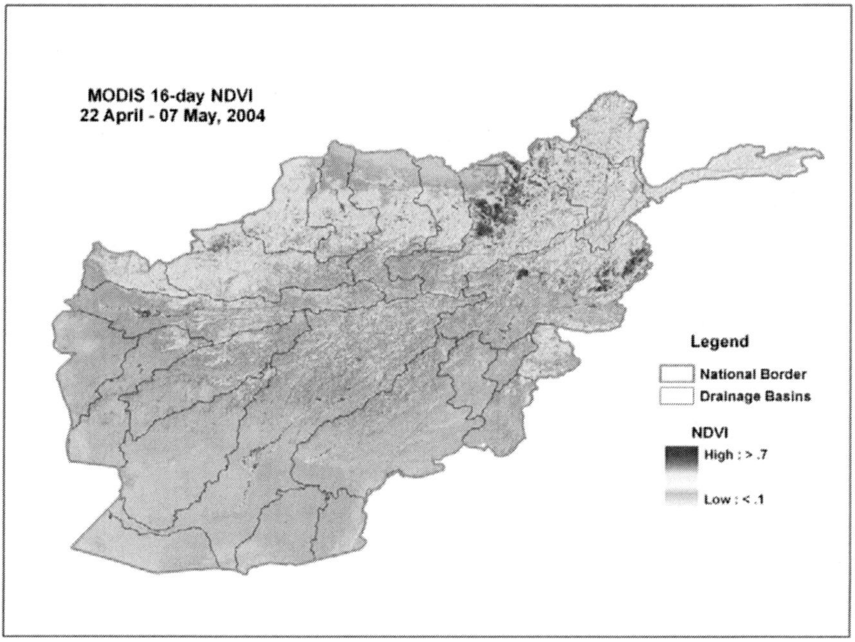

Figure 3. MODIS 250m 16-day NDVI composite in late April/early May 2004.

The MODIS vegetation index and land surface temperature data are distributed, at no cost, by the Land Processes Distributed Active Archive Center (LPDAAC), located at the USGS Center for Earth Resources Observation and Science (EROS) Center (http://LPDAAC.usgs.gov).

Reference ET: The global 1-degree reference ET (ETo), based on the 6-hourly Global Data Assimilation Systems (GDAS) model output, is calculated daily at EROS (Senay and Verdin [11]). The GDAS ETo (Figure 4) uses the standard Penman-Monteith equation as outlined in the FAO publication for short-grass ETo by Allen et al. [12]. The feasibility of using the GDAS ETo for such applications was recommended by Senay and Verdin [11] after a comparison with station-based daily ETo showed encouraging results with r^2 values exceeding 0.90. Daily global ETo values were available for all days between 2001 and 2005. For 2000, the daily ETo values were not complete. For the missing time periods, the average daily ETo from 2001 to 2005 was used.

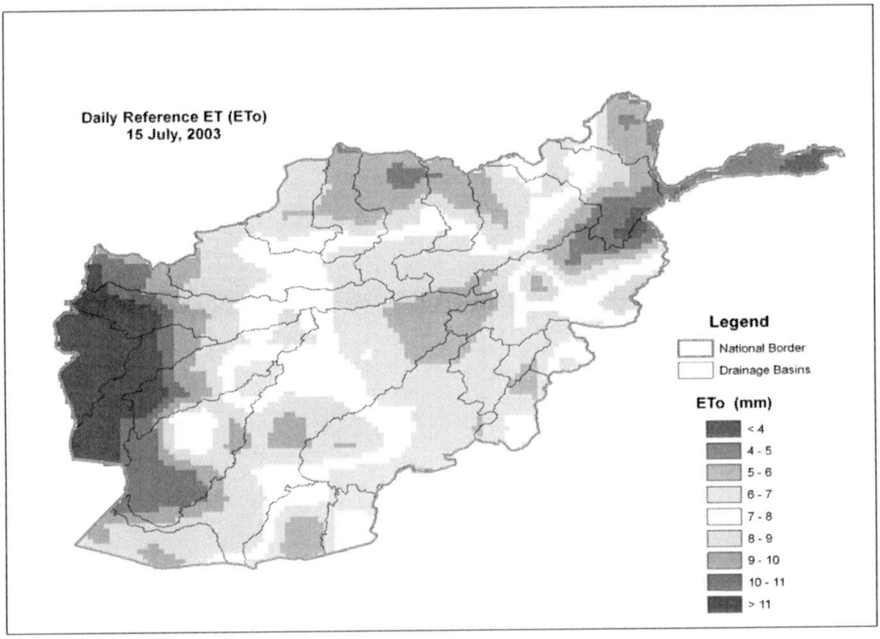

Figure 4. Daily GDAS reference ET (ETo) at 10km resolution, downscaled from 1 degree resolution using the IWMI long-term average gridded ETo.

Prior to using these data for our analysis, a downscaling of the GDAS 1-degree data was performed to produce a 10-km GDAS ETo. The approach utilized the International Water Management Institute (IWMI) historical (1961-90) potential evapotranspiration (PET). The IWMI data set is based on data originally compiled at 16-km spatial resolution that has been disaggregated to 10-km. The downscaling technique utilized the 10-km historical IWMI PET data to create a fractional relationship, on a per pixel basis, between each 10-km IWMI pixel (x) and the corresponding 100-km IWMI pixel (y) value (artificially created to simulate the 100-km GDAS). This fractional relationship was then applied back to the 100-km GDAS pixels to downscale these data to a spatial resolution of 10-km (Equation 2). The resulting 10-km ETo grids were used in this analysis (Figure 4). The downscaling method was developed and validated by the authors using similar IWMI data to develop the fractions over the conterminous US and the results compared favorably with station-based ETo in California. Thus, the method was proven to be an effective technique to downscale the Global GDAS ETo. A manuscript is being prepared to publish the validation of the downscaling technique.

$$x / y * \text{GDAS ETo} \tag{2}$$

Where, "x" represents 10-km IWMI PET and "y" is the 100-km IWMI PET

2.3. Procedures/Analysis

Sets of three hot and three cold pixels were selected for each 8-day composite period for each year of growing season data. An average of the 3 pixels was used to represent the hot and cold values in each of the six study areas. The study area was divided into 6 irrigation areas to reduce elevation differences that are large enough to cause land surface temperature differences since the premise of the existing model is based on the assumption that surface temperature differences are only caused by differences in moisture availability and water use. For a given time period, cold pixels, representing well vegetated and well watered crops, were selected based on a combination of low LST values and high values in the MODIS NDVI. Similarly, hot pixels, representing low density vegetation and relatively dry land, were identified by high LST values and very low NDVI values.

Land surface temperature values for each of the six pixels (3 hot, 3 cold) were extracted using ArcGIS 9.0 software [13]. The resulting database files were imported into an Excel spreadsheet where average hot and cold pixel values were calculated.

With the assumption that hot pixels experience very little ET and cold pixels represent maximum ET throughout the study area, the average temperature of hot and cold pixels could be used to calculate proportional fractions of ET on a per pixel basis. The ET fraction (ET_f) was calculated for each pixel by applying the following equation (Equation 3) to each of the 8-day MODIS land surface temperature grids.

$$ET_f = \frac{TH - Tx}{TH - TC} \tag{3}$$

where TH is the average of the three hot pixels selected for a given scene; TC is the average of the three cold pixels selected for that scene; and Tx is the land surface temperature value for any pixel in the composite scene.

The ET_f equation was applied to each 8-day average composite for each year, resulting in a series of more than 25 images per season. The images contained ET fractions for each pixel that were used to estimate actual ET throughout the growing season.

The ET_f is used in conjunction with ETo to calculate the per pixel actual ET (ETa) values in a given scene (Equation 4). The calculation procedure is similar to the use of a crop coefficient (Kc) and soil stress coefficient (Ks) in crop water-balance-based ET modeling where a ETo is adjusted in a multiplicative manner using Kc that is dependent on crop type and stage, and Ks which is dependent on soil moisture [1]. Daily ETo values were used to create average 8-day ETo corresponding to the LST composite periods.

$$Eta = ET_f * ETo \quad (4)$$

This simplified approach in SSEB allowed us to use modeled ETo at a coarse spatial resolution to derive spatially distributed high-resolution ETa values based on land surface temperature variability at 1-km resolution. Knowledge of ETa at a higher spatial resolution during the growing season provides important insight into the extent of irrigated cropland, the quality of the growing season, and associated seasonal water use.

Preliminary validation of the performance of SSEB was undertaken using crop fields in the north central United States. Corn and soybean fields in two South Dakota counties, Brookings and Moody, were used to compare model results with the SEBAL and METRIC methods. Landsat thermal scenes from August 1, 2000 and August 4, 2001 were used to assess the performance of the simplified energy balance method (SSEB). Depending on the crop type and year, the correlation coefficient (R^2) between SSEB and METRIC varied from 0.94 to 0.99 and with SEBAL from 0.55 to 0.79 (unpublished work by Swanson and Trooien, 2007, personal communication). A sample scatter plot of the validation work between SSEB and METRIC is shown in Figure 5.

The lack of field data in Afghanistan precludes a quantitative validation of the results in an absolute manner. However, quantitative analysis of the findings can be made by studying the year-to-year variability in relative terms. To this effect, actual crop ET for the six-year period, 2000 – 2005, was used to assess the quality of each growing season in the Helmand and Kabul study areas. Using masks of the irrigated crop area, described earlier in Figure 1, we used the 1-km gridded actual ET values to calculate spatially averaged actual ET for each season. The results and implications of these comparisons are outlined in detail in the following section.

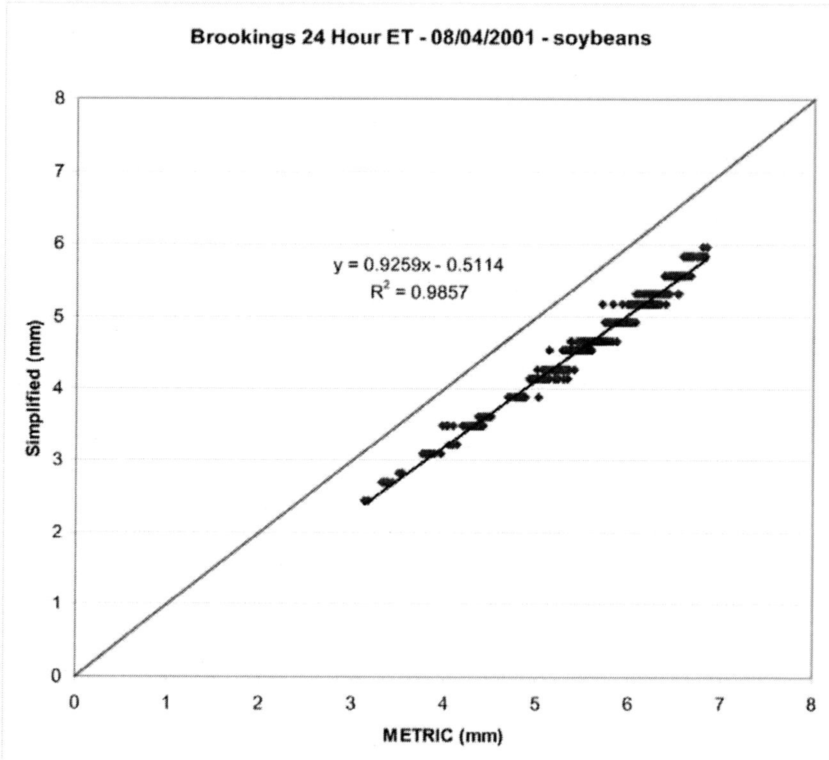

Figure 5. Scatter plot of one-day total actual ET (ETa) estimate between SSEB and METRIC model outputs in a soybean field on August 4, 2001. (Source: Swanson and Trooien, 2007)

3. RESULTS AND DISCUSSION

The presentation of results is divided into three primary sections. First, we provide a detailed discussion of the Kabul 1 area (Figure 1). Second, we will discuss in more general terms, the findings for all six study areas. Third, we will elaborate on further analysis that will be needed to provide a more comprehensive assessment of the seasonal irrigated crop demands throughout the Helmand and Kabul drainage basins.

3.1. Kabul 1

A time series of hot and cold pixel values for each 8-day time period in 2003 is shown in Figure 6. Similar temporal patterns were also observed in other years. For 2003, the hot and cold pixels were separated by an average of approximately 13 °C throughout the season between May and September. Furthermore, they appeared to increase or decrease in the same direction by about the same magnitude during the peak portions of the crop growing season. However, this separation approaches nearly to 0.0 °C during the off-seasons, particularly close to the start of the season. We believe the existence of such temporal patterns between the hot and cold pixels is potentially useful in detecting time of start of season for crop monitoring activities. However, early comparisons with seasonal NDVI time series indicate that the period of maximum water use corresponds more with atmospheric demand than with time of maximum vegetative stage, as measured using NDVI (Figure 7). Figure 6 also shows the boundary (extreme) conditions for surface temperature distribution in the study site.

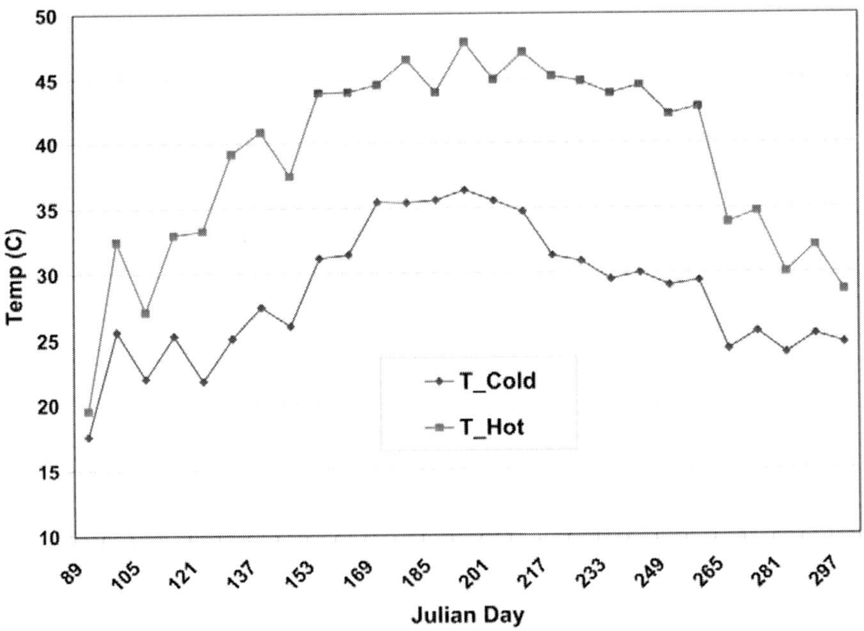

Figure 6. Temporal variation of the hot (T_hot) and cold (T_cold) pixel values during the 2003 crop growing season in °C.

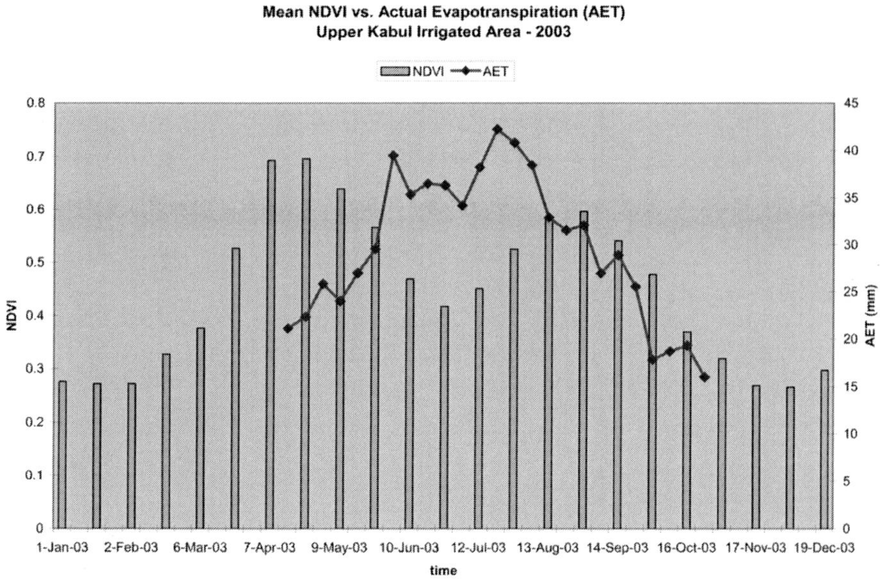

Figure 7. Temporal comparison of spatially averaged 16-day NDVI and 8-day ETa (AET) during the 2003 season.

By properly selecting the extreme temperature areas representing the hot (dry/bare) and the cold (wet/vegetated) land areas, the remaining pixels in the study area will fall in between these temperature values. The two extreme temperatures also correspond with extremes in ET values. The range of these values varies from zero ETa for the hot and dry areas to a high ETa, comparable in magnitude to ETo, for the cold and wet areas [3]. In this study, we extended this assumption to include the remaining pixels by suggesting that pixels having land surface temperature values in between these extremes will experience an ET value in direct proportion to the ET fraction as shown in equation 3.

Figure 8 shows the temporal trend of irrigation water use during the peak growing season for each of the 6 years used in this study. All years show a marked increase in seasonal water use in the late May and early June, a time when both air temperature and atmospheric demand are on the increase. The figure also shows a high level of inter-annual variability in seasonal water use, especially during the month of June. This could be explained by the fact that some crop fields may be taking advantage of spring rainfall and others are being harvested. Also, it may be that variability in spring rainfall could affect the timing and magnitude of seasonal snow melt, causing varying amounts of water to be available for irrigation use on different years. It should also be noted that the vegetation index temporal pattern identified in Figure 7 illustrates that this area is bimodal in its seasonality. Although Figure 7 shows only 2003 growing season NDVI, the bimodal pattern is seen throughout the six-year time series. The high degree of inter-annual variability seen in June is not evident later in the time series because irrigation water is not supplemented by other means. However, the 2003 growing season shows a distinct difference in irrigation water use compared to the other years. The irrigation water use from mid-July to early-August 2003 was 25 percent greater than the average. These findings corroborate field reports of an extensive second season crop during the 2003 growing season. All years showed a downward trend in water use beginning in the middle of August and progressing through September. This would correspond with a time when temperatures become milder and the difference between hot and cold pixels decreases.

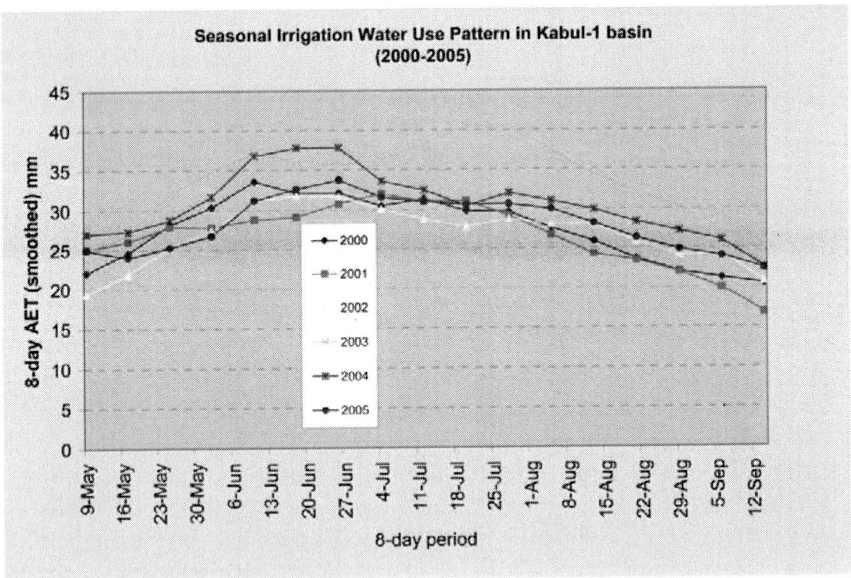

Figure 8. Temporal patterns of seasonal irrigation water use (ETa (AET)) in Kabul 1 for 2000 - 2005. Data is smoothed using a 3-period moving window.

Although the time series raises interesting questions about the temporal variability of water use patterns in the study area, an annual accounting of the irrigated water demand can be met by calculating the total seasonal ETa for each year. The actual crop evapotranspiration is estimated from the product of ET fraction and the ETo as shown by equation 4. Figure 8 shows the temporal patterns of the actual crop ET for each 8-day period. Each data point represents an 8-day actual ET estimate that is spatially averaged over the study area. The spatially-averaged seasonal ETa magnitudes are presented in Figure 9 to illustrate the year-to-year variability in actual ET.

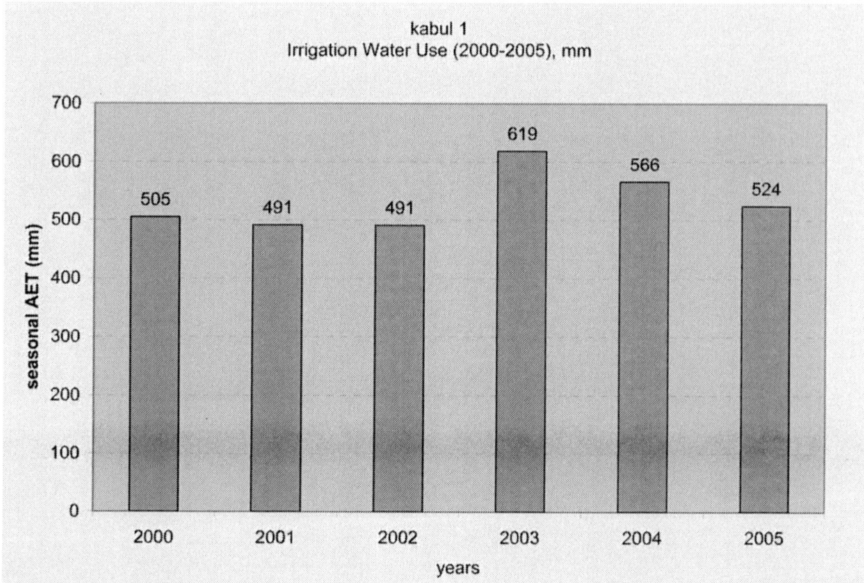

Figure 9. Spatially averaged peak-season (May – Sept.) actual crop ETa (AET) (mm) from 2000-2005.

Figure 9 highlights the fact that 2003 was the best agricultural season during the 5-year period, which is supported by several field reports that the cycle of three consecutive drought years, between 2000 and 2002, was alleviated by good precipitation in the 2003 season. These results are comparable to those from a watershed-based analysis of an operational FEWS NET irrigation supply and demand model, showing that 2002 and 2004 were below an average supply and the 2003 irrigation water supply met the average demand, defined using a 30-year rainfall climatology from 1961 to 1990 (www.cgiar.org/iwmi/WAtlas/atlas.htm). The spatially averaged seasonal ETa for 2003 is approximately 20 percent above the average of the other five years.

While Figures 7, 8 and 9 show spatially-averaged ETa, and Figure 10 presents the spatial variation of the seasonal actual ET in Kabul 1 for all six years. As is shown in Figure 9, the higher values of the 2003 seasonal ET are illustrated by the larger extent of high actual ET classes (darker green colors) compared to other years. Similar patterns of vegetation greenness for the corresponding years were observed from MODIS seasonal maximum NDVI data. The reduction in actual ET values, in all years except 2003, has mainly occurred between two

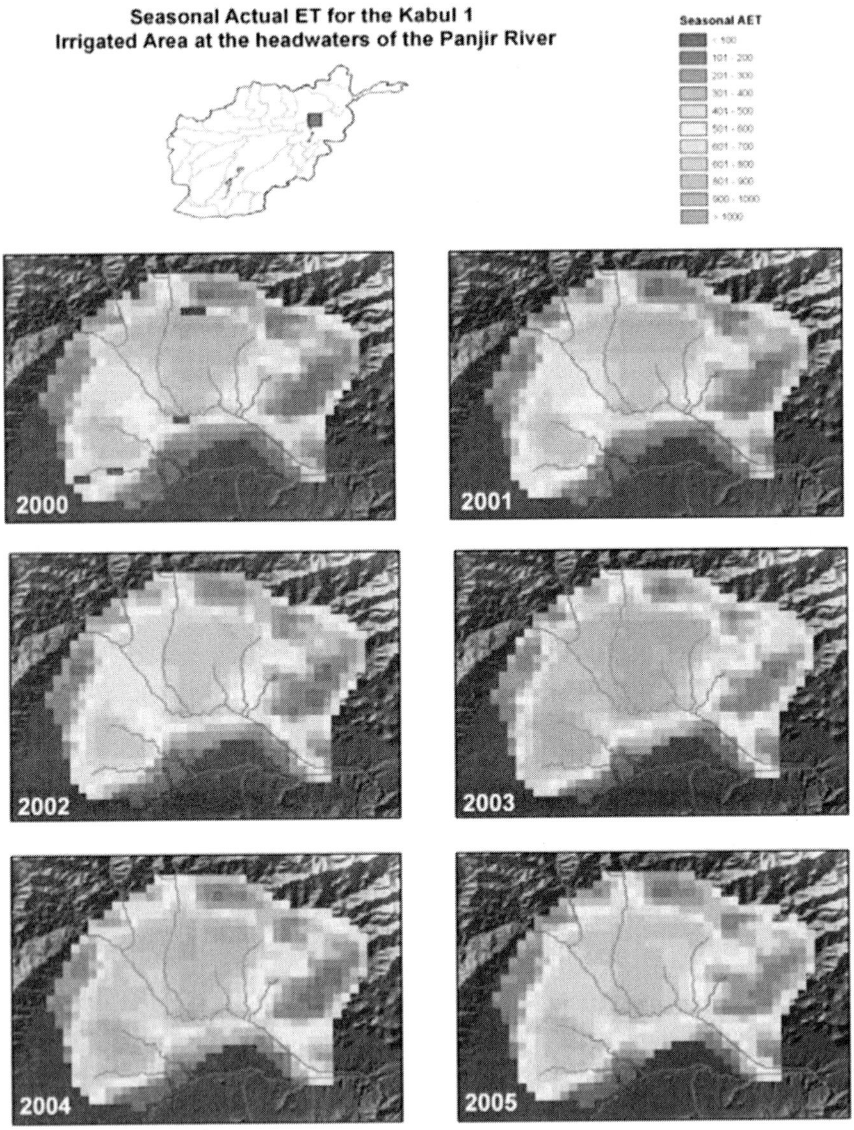

Figure 10. Seasonal (May – Sept.) actual ET (mm) distribution in irrigated fields of the Kabul 1 study area.

stream segments toward the southwest edge of the irrigated fields in Kabul 1, where a short tributary originates from a much lower elevation than the main stem to the north. The geographic area where lower actual ET values were observed seems to suggest that irrigators/fields further away from the major tributaries would have access to water only if there was a surplus in excess of the demands of the near-stream users. This is in line with a common practice in regions where water rights are not well established or regulated. Furthermore, Figure 10 suggests the possibility of using this method of analysis to estimate harvested irrigated areas for a given year based on a threshold of actual ET required for successful crop growth. For example, in the study area where certain fields consumed up to 800 mm in about 5 months, areas that only used half (400 mm) of the water demanded by a well-watered crop could be considered as unsuccessful and removed from harvested irrigated areas [1, 15]. Although the accuracy of the magnitudes of the estimated actual ET values requires field validation using other methods or field studies, the method's relative performance in terms of capturing the year-to-year variability suggests that the technique has a potential to characterize irrigated field crop performance in relative terms on reasonably homogeneous flat irrigated fields.

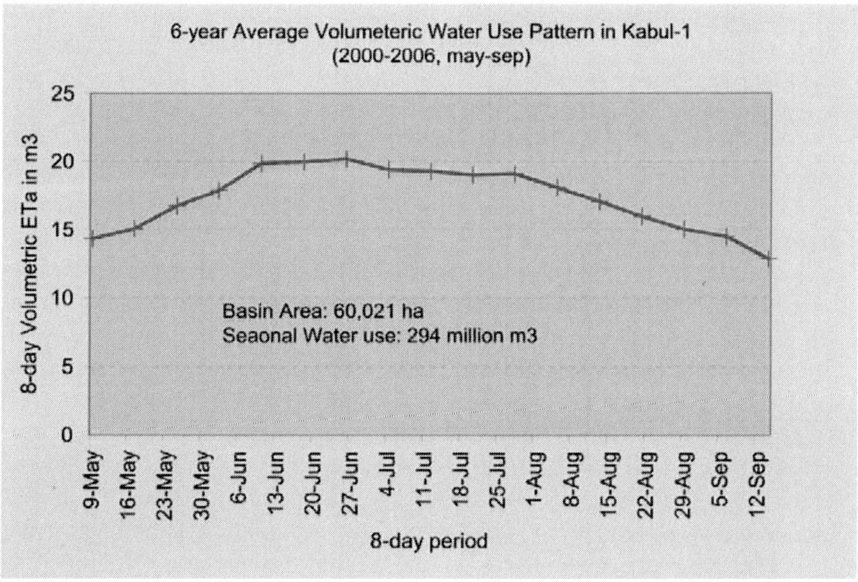

Figure 11. Estimated volumetric water use (ETa, m^3) pattern for the Kabul 1 irrigated area.

The next step in quantifying seasonal water use by irrigated crop land is to convert the ETa values into a volumetric assessment of water use. The method for doing this relies on the ability to calculate an irrigated basin area. Once the irrigated area is determined, the ETa values can be converted to a cubic measure of volumetric ETa. Figure 11 shows 8-day volumetric ETa values averaged over all six years of the study for May through September in Kabul 1. The result is a time series that allows for the assessment of temporal trends in volumetric water use as well as a quantification of seasonal volumetric water use throughout the basin. Preliminary calculations identify June as the month of highest volumetric water use and estimate average seasonal water use to be approximately 294 million cubic meters in Kabul 1. A full accounting of water use for the Kabul basin will require additional work as outlined in the further analysis section.

3.2. Other Study Areas

Similar detailed assessment of irrigated crop water use has been done for the other five basin sub-divisions. The graph shown in Figure 12 illustrates some of the similarities and differences in the findings for the other study sites. The most common feature among all areas is that the 2003 growing season stands out as having the highest seasonal ETa. Seasonal ETa of nearly 500 mm or greater was found for all study sites during the 2003 growing season. By contrast, almost every other year contained only one or two areas above 500 mm in seasonal ETa. Another characteristic among the six study sites was the high degree of inter-annual variability in seasonal ETa. However, the magnitude of this variability seems to show some regional differences. The Helmand basin areas appear to vary more dramatically than those in the Kabul basin. This seems to be somewhat counter-intuitive due to the fact that the upstream area of the Helmand basin contains the Kajaki Reservoir, which would be expected to provide a regulated release of water downstream. However, many of the irrigated areas of the Helmand stretch far beyond the source of irrigation water and therefore could be impacted by exploitation of water resources further upstream. As mentioned earlier, without proper regulations on water use and stream flow management, those irrigators nearer to the source may benefit the most while those further downstream are adequately supplied only in years of abundant water availability. Another possible explanation for the higher degree of variability in the Helmand region may be the result of crop rotation or fallow. A closer look at the Helmand 3 seasonal ETa (Figure 12) shows a fairly regular pattern of lower (300 – 400 mm) followed the next year by higher ETa (400 – 500 mm). It could be that some farmers in the region have fields that lay fallow every other year and therefore do not contribute to the seasonal ETa and consequently use less water on an inter-annual basis. Regional crop production reports may be able to provide additional insight into such cropping practices.

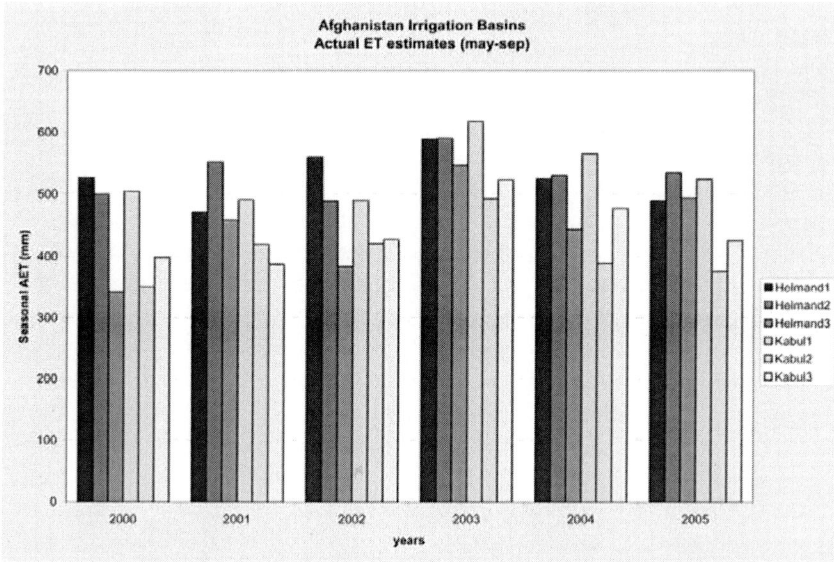

Figure 12. Seasonal ETa (AET) (mm) estimates (2000 – 2005) for all major irrigated areas in the Helmand and Kabul basins.

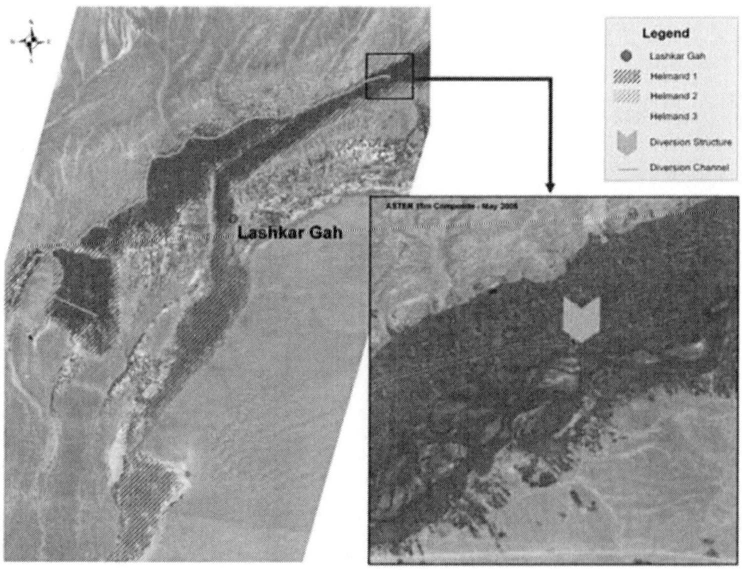

Figure 13. Location of a diversion structure on the Helmand River upstream of Lashkar Gah. The diversion channel is visible in the enlarged portion of the 15-m ASTER image above (right). The extent of the channel runs along the northern section of the Helmand 1 area, west of Lashkar Gah, and is highlighted in the image on the left.

Table 3 shows the relative water use variability among the 3 irrigated areas in the Helmand basin with respect to the location of a diversion structure upstream of the town of Lashkar Gah (Figure 13). The relative water use between the upstream area and the two downstream areas goes in the opposite direction from year to year. For example, when the upstream area (Helmand1) uses less water than the previous year (2000 versus 2001), the downstream areas use more water in the same season and vice-versa. The fact that an increase in water use by the upstream area is met with lower water uses in the downstream areas makes sense from the basic principle of water balance, demonstrating the capability of the model for capturing obvious water use patterns. On the other hand, the model appears to shed light on possible water management scenarios in the Helmand basin where high water use and low water use are alternating on a yearly basis for each of the basins. This could be a deliberate water allocation as regulated by the diversion structure upstream of Lashkar Gah or a process driven by production and market forces. Our speculation is based on the assumption that, for example, if the upstream basin produces well in one year, they may not need to produce as much the following year.

Table 3. Relative water use changes in the three irrigation areas of Helmand basin. Bracketed values are estimated seasonal actual ET (ETa) differences in mm. This table is a partial summary of graphical data presented in Figure 12.

Differencing Years	Helmand1 (upstream)	Helmand2 (downstream)	Helmand3 (downstream)
2001-2000	Decrease (-56)	Increase (+52)	Increase (+117)
2002-2001	Increase (+89)	Decrease (-63)	Decrease (-74)
2003*- anomalous	N/A	N/A	N/A
2006-2005	Decrease (-35)	Increase (+4 mm)	Increase (+51)

*: Since 2003 was an anomalously wet year, we eliminated (N/A) the use of 2003 data for this interpretation (i.e., 2004 minus 2003 and 2003 minus 2002).

4. DISCUSSION

Although the simplified surface energy balance (SSEB) approach has produced convincing results with encouraging comparisons with other established energy balance models, the model needs to be validated using actual field data. Further comparison with other more robust models will help identify its limitations and potential for different regions. Due to the prohibitive nature of collecting field data in this region of the world, we lack a good measure against which to validate our results. However, a recent comparison with the SEBAL model using Landsat Thermal data in China produced promising results with a coefficient of determination (R^2) of 0.89 (Cai Xueliang, 2007, personal communication). Figure 5 shows a similar comparison work in the US with the METRIC and SEBAL models where strong correlations have been exhibited. The authors intend to carry out a more detailed validation of the model in the United States using a combination of field measurements and results from water balance and other energy balance models. Currently, the model is appropriate to apply in a relatively homogeneous landscape where surface temperature variation is primarily caused by differences in water use. Therefore, major differences due to elevation, aspect, latitude, etc., should be taken into consideration before applying the method on the entire landscape. Furthermore, this method is somehow optimized to estimate actual ET (ETa) on crop surfaces since the albedo magnitude (0.23) embedded in the ETo is based on the grass reference ET. This is, however, a lesser problem than differences in elevation. Future improvement of the model will plan to take into account difference in albedo from other land cover types so that the model can be applied irrespective of land cover. In addition, the model may provide a high end estimate of the seasonal ET because it assumes the presence of well-watered, full-canopy, actively-growing vegetation during the early and late season of the crop cycle for the cold pixel. Again, further refinements of the model will include the use of NDVI to correct for vegetation cover condition to better estimate the actual ET at the wet pixel, based on a recommendation from Rick Allen (personal communication, 2006). Since an accurate estimation of the bare soil evaporation (Hot pixel) requires a water balance method, an integrated modeling framework that couples SSEB with a water-balance-based landscape ET model (VegET) is being developed at the Geographic Information Science Center of Excellence (GIScCE), a joint collaborative center between USGS/EROS and South Dakota State University.

5. CONCLUSIONS

The main objective of this study was achieved with the demonstration of the successful coupling of remotely sensed data and a simplified surface energy balance (SSEB) approach in producing estimates actual ET in irrigated agricultural areas of the Kabul and Helmand basins in Afghanistan. The ET fractions are comparable to the commonly used crop coefficient (Kc) (with a soil water stress factor, Ks) and thus are directly multiplied with ETo values to obtain actual ET estimates. A major advantage of the energy-balance approach is that it can be used to quantify spatial extent of irrigated fields and their water-use dynamics without a reference to source of water as opposed to a water-balance model which requires knowledge of both the magnitude and temporal distribution of rainfall and irrigation applied to fields.

The actual ET values of the irrigated fields in the study area varied from year to year in a way that was consistent with field reports and other independent data sets such as the seasonal maximum NDVI and output from an irrigation supply/demand water balance model. The 2003 seasonal actual ET of the study area was much higher than the rest of the studied years, and about 20 percent higher than the average of the 5 years. This was in agreement with published reports [14] and news sources stating that 2003 precipitation appeared to have broken the sequence of preceding consecutive dry years in Afghanistan.

A close examination of the spatial distribution of the actual ET estimates during the six years revealed that the reduction in area of high actual ET values during less intense water use years was primarily in an area further from the main stream channels. Because this corresponds to a common practice in which water is generally used first by those upstream, the result reinforces the reliability of this approach and points to the potential application of this method for spatially estimating cropped area in irrigated fields. The method also captured probable water management scenarios where a unique year-to-year variability was identified in addition to water use difference between upstream and downstream areas. The existence of a variable temporal pattern between the hot and cold pixels during the crop season is also believed to be potentially useful in detecting time of start of season for crop monitoring activities. However, further examination of the relationship between time series of ETa and NDVI is needed.

Although the results obtained from the thermal-based ET fraction approach in SSEB are encouraging for applications in remote locations where field-based information is not readily available, it is recommended that the method needs to be further investigated and validated using more detailed surface energy balance methods and field data under different hydro-climatic conditions.

REFERENCES

1. Senay, G.B.; Verdin, J. Characterization of Yield Reduction in Ethiopia Using a GIS-Based Crop Water Balance Model. Canadian Journal of Remote Sensing **2003**, 29(6), 687–692.
2. Verdin, J.; Klaver, R. Grid cell based crop water accounting for the famine early warning system. Hydrological Processes **2002**, 16, 1617–1630.
3. Allen, R.G.; Tasurmi, M.; Morse, A.T.; Trezza, R. A Landsat-based Energy Balance and Evapotranspiration Model in Western US Water Rights Regulation and Planning. Journal of Irrigation and Drainage Systems **2005**, 19(3-4), 251–268, 18.

4. Bastiaanssen, W.G.M.; Menenti, M.; Feddes, R.A.; Holtslag, A.A.M. A remote sensing surface energy balance algorithm for land (SEBAL): 1) Formulation. Journal of Hydrology **1998**, 212(213), 213–229.
5. Bastiaanssen, W.G.M. SEBAL-based sensible and latent heat fluxes in the irrigated Gediz Basin, Turkey. Journal of Hydrology **2000**, 229, 87–100.
6. Bastiaanssen, W.G.M.; Noordman, E.J.M.; Pelgrum, H.; Davids, G.; Thoreson, B.P.; Allen, R.G. SEBALmodel with remotely sensed data to improve water resources management under actual field conditions. Journal of irrigation and drainage engineering **2005**, 131(1), 85–93.
7. Su, H.; McCabe, M.F.; Wood, E. F.; Su, Z.; Prueger, J. Modeling evapotranspiration during SMACEX: comparing two approaches for local and regional scale prediction. Journal of Hydrometeorology **2005**, 6(6), 910–922.
8. Sadler, E. J.; Bauer, P. J.; Busscher, W.J.; Millen, J. A. Site-specific analysis of a droughted corn crop: water use and stress. Agronomy Journal **2000**, 92, 403–410.
9. Jackson, R.D. Canopy temperature and crop water stress. Vol 1, In Advances in irrigation.; Academic Press: New York, 1982; pp. 43–85.
10. Qiu, G. Y.; Miyamoto, K.; Sase, S.; Okushima, L. Detection of Crop Transpiration and Water Stress by Temperature Related Approach Under Field and Greenhouse Conditions; Department of Land Improvement, National Research Institute of Agricultural Engineering: Tsukuba, Ibraki, Japan, 1999.
11. Senay, G.B.; Verdin, J. Evaluating reference evapotranspiration (ETo) model output from the Global Data Assimilation System using station ETo in the US. Proceedings of Reclamation ET workshop: state-of-the-art review of ET remote sensing science and technology, Fort Collins, Colorado, Feb 8-10, 2005.
12. Allen, R.G.; Pereira, L.; Raes, D.; Smith, M. Crop Evapotranspiration.; Food and Agriculture Organization of the United Nations: Rome, Italy, 1998; ISBN: ISBN 92-5-104219-5. p. 290p.
13. ESRI. ArcGIS 9.0; Environmental Systems Research Institute (ESRI): Redlands, CA, 2004.
14. FEWS NET. Afghanistan Monthly Food Security Bulletin. September 2003. www.fews.net.
15. Doorenbos, J.; Pruitt, W.O. Crop water requirements.; FAO Irrigation and Drainage Paper No. 24; Food and Agricultural Organization of the United Nations: Rome, Italy, 1977; p. 144p.

Index

C

Cambodia, 113, 114, 115, 116, 120, 125, 127, 128, 130, 131, 132, 133

Cotton Seedling, 17, 18, 28, 36, 37

crop coefficient, 62, 63, 64, 82, 94, 99, 100, 101, 102, 103, 165, 180, 248, 259

E

energy balance, 58, 62, 86, 113, 114, 118, 119, 127, 130, 137, 259, 260, 261

Energy Balance Bowen Ratio Method, 114, 118

ETMonitor, 181, 182, 183, 185, 202, 203, 204, 209, 211

F

Field Measurement, 17, 18

G

Greenhouse Gas Emissions, 222

H

Heihe River basin, 181, 182, 183, 184, 185, 192, 195, 196, 198, 200, 203, 204, 205, 206, 207, 208, 210, 211, 216

Hydraulic Load, 172

Hydrological Modeling, 90

I

Irrigation Agriculture, 2

L

Landfills, 221, 222, 234, 235, 236

Landsat 8, 61, 62, 65, 66, 70, 71, 75, 76

Leaf Area Index, 26, 66, 70, 189

low permeability, 153, 154, 155, 177, 180

Lowland Dry Evergreen Forest, 113, 114

Lysimeter, 18, 37, 38

M

Methane Oxidation, 222, 235, 236
Microclimate, 85, 135, 140
Micro-Lysimeters, 23
Multilayered Model, 114

O

on-site wastewater, 154, 176, 179, 180

P

passive treatment, 154, 177

R

RapidEye, 61, 62, 65, 66, 70, 71, 74, 75
reference ET, 238, 240, 242, 245, 246, 259
Reference Evapotranspiration, 2, 16, 110, 111
remote sensing, 58, 62, 63, 76, 83, 84, 136, 181, 182, 183, 185, 192, 193, 196, 202, 203, 210, 212, 213, 214, 216, 218, 219, 220, 261

S

Soil Porosity, 158
spatial and temporal variations, 182, 204
SSEB, 238, 240, 248, 249, 258, 259, 260
Sugarcane, 89, 90, 100, 107, 108, 109, 111
SWAT, 89, 90, 92, 96, 99, 102, 103, 107

T

table grapes, 61, 62, 63, 66, 72, 76, 77, 78, 80, 81, 84, 87
Transpiration, 18, 38, 83, 131, 136, 261
Turbulent Transport, 135

V

vineyards, 61, 62, 63, 64, 65, 70, 76, 77, 79, 82, 85, 87

W

Water Balance, 15, 39, 89, 90, 103, 108, 161, 221, 234, 260
willow, 153, 154, 155, 157, 158, 172, 173, 174, 175, 176, 177, 178, 179, 180